DEMOCRACY AND TECHNOLOGY

THE CONDUCT OF SCIENCE SERIES

Steve Fuller, Ph.D., Editor
DEPARTMENT OF SOCIOLOGY
UNIVERSITY OF DURHAM, ENGLAND

Democracy and Technology
RICHARD E. SCLOVE

The Social Psychology of Science
WILLIAM R. SHADISH AND STEVE FULLER

Philosophy of Science and Its Discontents, Second Edition
STEVE FULLER

The Scientific Attitude, Second Edition
FREDERICK GRINNELL

Politics and Technology
JOHN STREET

DEMOCRACY AND TECHNOLOGY

Richard E. Sclove

THE GUILFORD PRESS
New York London

© 1995 The Guilford Press
A Division of Guilford Publications, Inc.
72 Spring Street, New York, NY 10012

Printed in the United States of America

This book is printed on acid-free paper.

Last digit is print number: 9 8 7 6 5 4 3 2

Library of Congress Cataloging-in-Publication Data

Sclove, Richard.
    Democracy and technology / Richard E. Sclove
      p. cm. — (Conduct of science series)
    Includes bibliographical references and index.
    ISBN 0-89862-860-1. — ISBN 0-89862-861-X (pbk.)
    1. Political science. 2. Technology—Political aspects.
    3. Technology—Social aspects. 4. Democracy. 5. Political
planning. 6. Policy sciences I. Title. II. Series.
JA80 . S58 1995
303.48'3—dc20   95-18005
                                                 CIP

CONTENTS

LIST OF FIGURES

PREFACE

This book promotes the reconstruction of technology along more democratic lines. Like many books, it germinated in response to personal questions and challenges. When I was 24, I landed a job as the research assistant to a major energy policy study. The job was a plum. The senior study team included a bevy of Harvard professors, a Nobel Prize winner, a director of the World Bank, and several celebrated scientists. I attended working dinners with congressmen and senators, and conferences in Aspen, Bermuda, and Paris.

For me, there was just one problem. I gradually grew convinced that conventional approaches to policy analysis obscure many important ways that energy and other technologies can transform people's lives. I was also troubled by a sense that even the most well-intentioned, elite study group can be deeply unaware of the extent to which its conclusions embody far-reaching value judgments. The problem was not at all in how our group applied economic analysis and its other analytic methods, for the study team was dazzlingly skilled. The problem lay somewhere in the methods themselves, as well as in the social processes through which they were implemented.

These are the distant personal origins of this volume—my own quest to discover whether there are better ways to make decisions about the technologies that seem almost daily to make, unmake, and remake our world. My basic argument is simple: Insofar as (1) citizens ought to be empowered to participate in shaping their society's basic circumstances and (2) technologies profoundly affect and partly constitute those circumstances, it follows that (3) technological design and practice should be democratized.

The resulting work combines knowledge from many disciplines. It is intended primarily for anyone interested in democracy and public policy, social justice and empowerment, political economy and business, or the social consequences of technology and architecture. The book may also prove methodologically helpful to social scientists, historians, and philosophers, including those who have not previously explored the complex role of technology within social systems.

I began this work more than a decade ago, during the early 1980s—a time when U.S. technology policymaking was dominated by the Reagan administration's Cold war military preoccupations. To envision democratizing technological decisions and design under those circumstances required something of a heroic leap of faith or, at the very least, an exceptionally evolved readiness to delay gratification.

In the meantime, the world has changed in ways that were then inconceivable. For a half-century the Cold war provided the overarching setting within which U.S. science and technology policies were formulated. Thus the Cold war's sudden collapse has dissolved the dominant rationale undergirding many policies and institutions. This establishes a strategic opportunity to make technology decision making and design more responsive to democratically decided social needs. Under these circumstances the requisite leap of faith diminishes appreciably.

Likewise, elsewhere in the world—in the European Community, among former Warsaw Pact nations, and in the developing world—the turn to more open or internationally integrated trade regimes, laissez-faire capitalism, or shock capitalism is inevitably producing anxieties, disappointments, and in some places catastrophes of its own. This, together with new global movements toward democracy, may also signal new opportunities and needs for creative approaches to the interrelations of technology, society, and politics.

The book's argument unfolds in three stages. Part I synthesizes two disparate bodies of knowledge. One is a corpus of recent research into the social dimensions of technology. The other is that body of knowledge and practice known as democratic theory. These domains of knowledge have only rarely been related to one another, but their combination swiftly issues in the rudiments of a comprehensive democratic theory of technology.

Part II develops a provisional system of design criteria for distinguishing technologies that are compatible with democracy from those that are not. This undertaking is unusual, for there is little precedent for using political philosophy to develop prescriptions for technological design and choice.

Part III elaborates the concept of a democratic politics of technology. Challenging the foundations of modern economic thought, I argue that the democratic theory of technology qualifies as a coherent alternative to neoclassical welfare economics. Indeed, reinvigorated democratic politics should largely supersede conventional economic reasoning as a basis for technological decisions. I review case studies of participation by laypeople in technological research, development, and design. Finally, I explore political steps and strategies that can help us achieve citizenship in a future world of democratic technology. That world—so unlike today's and yet within our reach—would witness technological evolution becoming subordinate to democratic prerogatives. It would be a world made *by* people but also *for* people, acting under circumstances more favorable to fair and informed results.

ACKNOWLEDGMENTS

My quest toward a democratic theory of technology was generously assisted by many good people. For commenting on portions of the book or its precursors, I wish to thank Hayward Alker, Brian Baker, Andy Belser, Michael Black, Iain Boal, Eric Brende, Larry Bucciarelli, Vary Coates, Joshua Cohen, Betsy Conyers, Susan Cozzens, Hubert Dreyfus, Paul Durbin, David Elliott, Steve Fuller, Herb Gintis, Michael Goldhaber, Peter Haas, Patrick Hamlett, William W. Hogan, Don Ihde, Martin Kessler, Hans Klein, Allan Krass, Todd R. LaPorte, Larry Lidsky, Genevieve MacLellan, Laura Nader, David F. Noble, Dick Norgaard, George Perkovich, Wolf Schäfer, Jeff Scheuer, Seth Shulman, Michael Shuman, Frank von Hippel, Charlie Weiner, Woody Wickham, Langdon Winner, Ned Woodhouse, Rick Worthington, the late Hugh Aitken, and two constructive anonymous reviewers.

I also benefited from thoughtful conversations with Peter Buck, Michel Callon, Gary Chapman, Mike Cooley, Clem Dinsmore, Martin Diskin, Colleen Dunlavy, Frank Emspak, Richard Fink, Lisa Greber, Dan Grossman, George Hoberg, Frank Laird, Bruno Latour, Heather Lechtman, Peter Lipton, Jerry Mander, Leo Marx, Carl Mitcham, Guild Nichols, David Orr, Marcus Raskin, Sal Restivo, Charley Richardson, Geoffrey Sea, Michael Shannon, Eugene Skolnikoff, Sandra Tanenbaum, Robert H. Williams, and Joel Yudken.

My research was made possible in part through the award of a Ciriacy–Wantrup Postdoctoral Fellowship in Resource Economics at the University of California at Berkeley and a Copeland Fellowship at Amherst College, as well as through the generosity of the Menemsha Fund, Rockefeller Family Associates, The Foundation for Deep Ecology,

The John D. and Catherine T. MacArthur Foundation, and Harvard University's Energy and Environmental Policy Center. I was graciously hosted at various stages by the Center for International Studies at M.I.T., the Institute of Governmental Studies at the University of California at Berkeley, Amherst College's Chemistry Department, Katz's B&B, and the Institute for Policy Studies in Washington, DC. At the University of Massachusetts at Amherst I received warm hospitality and assistance from the Economics Department, the Institute for Advanced Study in the Humanities, and the Program in Science, Technology, and Society. The staffs of Amherst College's library and academic computer center also provided invaluable support.

My wife, Marcie Abramson Sclove, supported me in untold ways, not only with her love but also through her ever-ready willingness to hear me out and set me straight. She did not, however, sacrifice her own life to do so, as her many independent accomplishments will attest. Our daughter, Lena, always possessed of good humor and sense, chose wisely to appear after the bulk of the work was done. Those who know them understand my remarkable good fortune in having these two people in my life.

Thank you one and all.

The Loka Institute
Amherst, Massachusetts

PART I

The Nuts and Bolts of Democracy

Chapter 1

SPANISH WATERS, AMISH FARMING
Two Parables of Modernity?

> I wish to . . . persuade those who are
> concerned with maintaining
> democratic institutions to see that
> their constructive efforts must
> include technology itself.
> —*Lewis Mumford*[1]

During the early 1970s, running water was installed in the houses
of Ibieca, a small village in northeast Spain. With pipes running directly
to their homes, Ibiecans no longer had to fetch water from the village
fountain. Families gradually purchased washing machines, and women
stopped gathering to scrub laundry by hand at the village washbasin.

Arduous tasks were rendered technologically superfluous, but village social life unexpectedly changed. The public fountain and washbasin, once scenes of vigorous social interaction, became nearly de-
serted. Men began losing their sense of familiarity with the children and
the donkeys that had once helped them to haul water. Women stopped
congregating at the washbasin to intermix their scrubbing with politically empowering gossip about men and village life. In hindsight, the
installation of running water helped break down the Ibiecans' strong
bonds—with one another, with their animals, and with the land—that
had knit them together as a community.[2]

Is this a parable for our time? Like Ibiecans, we acquiesce in
seemingly innocuous technological changes.* Unlike many of the Ibie-

---

*"We" in this sentence is not a royal plural. Throughout the text "we" means, depending
on the context, either me (the author) and you (the reader) or else contemporary citizens
generally.

cans, however, we celebrate these changes: whiter teeth, lower cost or else greater convenience, abundance, safety, or amusement. The automobile, for example, embodies a distinctively American conception of freedom. People speed through city and countryside toward adventure and opportunity. But the negative results of our many individual decisions to purchase automobiles include gridlock, air pollution, suburban sprawl, the decline of downtown centers, and dependence on insecure sources of imported oil. Did we choose these results? Do they express people's freedom, or perhaps ironically limit it?

Of course, the automobile's adverse effects were never intended, any more than Ibiecans intended to dissolve their former way of life by introducing running water. Ibiecans did not foresee the extent to which earning money to own a washing machine would mean becoming enmeshed in the external cash economy. They did not plan to remake themselves into wage laborers and consumers, nor did they plan to gradually transform their town into the suburban appendage of an encroaching urban center. For many Ibiecans, the loss of their old way of life and the consequent pain proved profound. One farmer, compelled to sell his beloved but now useless donkey, withered into permanent silence. For Ibiecans, as for everyone else, the combined result of many individual technological choices is often not what anyone anticipated.

Modern industrial nations have, of course, outdone rural villages in evolving social processes for coping with technologies' unwanted effects. On the one hand, a modest scholarly industry, steeped in economic ideas, stumps for policies to accelerate technological innovation.[3] Their objective is to enhance national economic growth, productivity, and international competitiveness, based on the assumption that as long as an innovation sells profitably, it is a social blessing. But newspapers also grant front-page coverage to controversial technological developments—to industrial disasters or to unsettling advances in genetic engineering, automation, and weaponry. The United States has an Environmental Protection Agency to regulate technologies' impact on the environment. The Occupational Safety and Health Administration is responsible for worker safety, the Defense Department supports innovations in military hardware, many government agencies forecast technological trends, Congress provides oversight and legislative initiatives, the courts offer redress for grievances, and various private and nonprofit groups strive to advance their views of the public good.

But here, too, something is missing, something so vast that it is easy to overlook: virtually the entire range of technologies' psychological, cultural, and political effects.[4] For example, newspapers, public-interest groups, corporate leaders and governmental bodies—when they consider a technology—normally address one or more of the following four questions: Is the technology at issue technically workable? What are its economic costs and benefits, and how are they distributed? What are the associated environmental, health, and safety risks? Are there implications for national security?

Undeniably, these are important questions. Yet as a group they are incomplete, for they fail to address technologies' profound role in altering the course of history and the texture of daily life. Consider the difference it would have made to us today had our forebears learned to pose these questions—and then act responsibly on the answers—throughout the first century and a half of industrialization. The world today would be cleaner and safer; thus, in certain significant respects, we would be better off. However, our societies would still have done nothing directly to comprehend, not to mention to guide or perhaps to alter or avert, such major, technologically influenced developments as establishing the home as a place where a woman labored alone, the birth of the nuclear family, changing sexual mores, suburbanization, the development of public schools and romanticized childhood, the withering of craftsmanship, the shift from an agrarian/cyclic experience of time to a linear one, the creation of hierarchically managed national and transnational corporations, or the evolution of modern political parties.[5]

In short, with attention confined strictly to these four questions, the momentous cultural developments associated with the Industrial Revolution would have come and gone without anyone noticing. Yet these questions, which are incapable even of distinguishing actions that perpetuate an agrarian social order from those that promote revolutionary political and cultural transformation, are the very questions now imagined adequate to guide us wisely into the next century.

This complicity in technological decisions that haphazardly uproot established ways of life is as perplexing as discovering a family that shared its home with a temperamental elephant, and yet never discussed—somehow did not even notice—the beast's pervasive influence on every facet of their lives. It is even as though everyone in a nation were to gather together nightly in their dreams—assemble solemnly in a glistening moonlit glade—and there debate and ratify a new constitution. Awakening afterward with no memory of what had passed, they nonetheless mysteriously comply with the nocturnally revolutionized document in its every word and letter. Such a world, in which unconscious collective actions govern waking reality, is the world that now exists.[6] It is the modern technological world that we have all helped create.

Could it be otherwise? Are the social effects of technology truly so complex that no one could possibly foresee them, much less act cogently to guide them? Not necessarily. To demonstrate this, one can contrast both Ibieca's and contemporary American society's style of technological politics with that of an alternative social order.

The Old Order Amish immigrated to the United States during the 18th and 19th centuries. With established communities in some 20 American states, their U.S. population is more than 100,000 and growing. To the outside society they are known as a religious subculture distinguished by old-fashioned clothing, horse-and-buggy transport, and an antiquated lifestyle that rejects modern technologies. The actual story is more complex and instructive.

The Amish are a pragmatic people who accept the reality of social change and do not reject all modern technology. Hence, theirs is not a primitive folk culture that lacks awareness of alternative possibilities. On the contrary, they represent a society that is conscious of the larger world in which it is immersed and that self-consciously guides its evolution.[7] The Amish have, for example, repeatedly adopted innovations in farming technology, sometimes sooner than their non-Amish neighbors. They will hitch a ride in a non-Amish car, charter a bus and driver, or perform sums using a battery-powered hand calculator. They are also skilled technological innovators who have been known, for instance, to devise a system in which a diesel tractor powers an air compressor that, in turn, pumps refined fuel to a set of indoor lamps. On the other hand, most Amish communities forbid personal ownership of automobiles, telephones, radios, and televisions; the use of tractors in the field; and electric hookups from power company grids to private homes and buildings.

To a casual observer, the resulting pattern of exclusions and adoptions seems capricious. However, the pattern is the result of a remarkably sophisticated style of technological politics. The exact decision-making process varies somewhat from one Amish community to the next and from one decision to the next. In essence, each local Amish community—acting collectively rather than as a set of discrete individuals—asks itself how the adoption of a technology would affect the community as a whole. Innovations that would tend, on balance, to preserve the community, its religion, and its harmonious relation with nature are permitted; those that appear to threaten the community and its values are rejected. In either case, the decision is reached through a process of public discussion and democratic ratification.[8]

What would be the impact on our desired form of society if individuals, or the community, were to adopt one set of technologies rather than another? The villagers of Ibieca had no tradition of asking such questions or even an established forum for making the attempt. Nor do we. But isn't it striking that the Amish, who prohibit formal schooling past the eighth grade, have nevertheless managed for several centuries to make technological choices that shrewdly advance their chosen cultural and religious commitments? In this regard, their technological acumen surpasses that of the villagers of Ibieca as well as the combined capability of modern nations' scientific, commercial, and policymaking establishments.

Reconsider, then, our society's ineptness at guiding technological change. Might it have less to do with modern technological complexity than with a failure to evolve institutions through which we could begin to act upon appropriate questions? The potential list of neglected questions concerning technology is long. It could encompass the entire domain of technologies' social aspect: the political, cultural, sociological, psychological, and spiritual realms. Moreover, one might need to integrate such issues with others more familiar—matters of technical feasibility, economics, environment, health, and defense. Finally, it might be necessary to consider not only the social dimensions and impacts associated with single technologies, but also the combined effects that emerge from a complex of coexisting technologies.

Were we to do this, it might emerge that technologies, everyday tools and helpers, are implicated in a plethora of modern ills, including loneliness, narcissism, disempowerment, insecurity, stress, and alienation. Stated more concretely, technology appears to contribute indirectly to problems ranging from urban poverty to teenage pregnancy, child abuse, racism, the continued subordination of women, militarism, the marginalization of the elderly, high crime rates, and drug abuse. Ultimately, technology is implicated in perpetuating antidemocratic power relations and in eroding social contexts for developing and expressing citizenship.

Technology is not *the* cause of such ills, but it contributes to all of them. To continue to neglect technologies' broad social dimensions virtually guarantees that we will remain ineffectual in addressing our deepest social problems and sources of personal malaise. It will not do, moreover, to imagine that other kinds of social reforms—be they conservative or radical—must precede significant reform in the technological domain, such that we must "First transform society, then tackle technology." That refrain overlooks ways that existing technologies help constitute the present social order and so constrain social transformation. Until technological concerns are fully integrated into programs of social transformation, such programs will be stunted or abortive.

Several qualifications are in order. First, insofar as technology is not the sole contributor to social problems, one ought not to shift attention to technology at the expense of other contributing factors. Concentrated economic power, poverty, racism, sexism, ethnic intolerance, and so on matter too; it is thus vital to explore the relationship between technology and these other factors.

Second, it is wrong to conclude that "technology is evil; let's get rid of it." We can no more eliminate technology than we can cease to be human. However, third, neither must we merely adapt compliantly to whatever technologies happen along. An adequate approach to technology must involve procedures for addressing a broader, more appropriate set of questions. But these must lead to the possibility of eliciting alternative technologies more compatible with the kind of society or communities in which people wish to live.

Among the panoply of questions concerning technology that escape attention, perhaps the most important one involves how technology bears on democracy. Democracy provides the precondition for being able to decide fairly and effectively what further questions to ask and what actions to take in light of the answers. Thus if technologies were more compatible with one or another vigorous variant of democracy, we might be better positioned to debate what other issues most urgently require attention. Conversely, it is vital to explore the extent to which the failure to come to terms with technologies' political ramifications represents an expression of antidemocratic social power formations, as embodied partly in current technologies themselves.

For a preliminary illustration of the importance of seeking compatibility between technology and democracy, let us turn again to the Amish. The Old Order Amish ask themselves how a particular set of technologies would affect their community. However, it happens that their communities already embody a relatively robust species of local, democratic self-governance.[9] Hence, implicit in the question of how to preserve their community is the crucial subsidiary concern that any permitted technologies must be compatible with preserving the Amish community's already-democratic nature.

The Amish have, for instance, prohibited private ownership of automobiles. This is done in part to inhibit a dispersed settlement pattern that would interfere with Amish-style extended families and neighborliness.[10] Such neighborliness is pleasurable and also necessary to promote economic mutuality and to perpetuate Amish culture. Furthermore, it contributes to the kind of mutual understanding, social commitment, and routine of gathering that, in turn, facilitate participatory and consensual decision making. Were the Amish to purchase automobiles, they would be jeopardizing their ability to continue governing themselves democratically with respect to technology and otherwise.

This does not mean that everyone should become Amish or impulsively discard their automobiles. Nor should one overlook features, such as smallness and cultural homogeneity, that distinguish Amish society from the U.S. mainstream. It is doubtless easier for the Amish to achieve consensual decisions than it would be for the citizens of a large, culturally diverse city. But for immediate purposes, the problem of achieving consensual answers is of much less concern than our failure even to begin debating crucial questions—in this case, concerning technologies' political and cultural dimensions.

In short, the "nuts and bolts of democracy"—ordinarily a metaphor denoting concern with the nitty-gritty of democratic politics—must grow to encompass a literal concern with nuts and bolts. Currently, there are few institutions through which citizens can become critically engaged with choosing or designing technologies. Should we commit ourselves to evolving such institutions and to adopting only those technologies that are compatible with democracy? Until we do, I shall argue, there can be no democracy worthy of the name.

Chapter 2

I'D HAMMER OUT FREEDOM
Technology as Politics and Culture

> Technological innovations are
> similar to legislative acts or political
> foundings that establish a
> framework for public order that will
> endure over many generations.
> —*Langdon Winner*[1]

What is technology? People ordinarily think of technology as machinery or gadgetry, as an economic factor of production, as know-how, as what engineers do, or as progress. Often they characterize technologies in terms of a single intended function. What is a hammer? It's what someone uses to pound nails into boards. What is a telephone? It's a device that enables people to converse at a distance. Some technologies, however, have more than one intended function. Hammers, for example, can pound nails into boards but can also extract them. This is the core of the contemporary view of technology. People understand technologies in terms of a primary function—or, occasionally, several functions—that each is intended to accomplish.

Beyond this, our society has in the past few decades come to acknowledge that technologies tend to produce at least two general kinds of "secondary" or "unintended" effects. First, they generate environmental consequences: pollution, resource depletion, and ecosystem modification. Each of these may, in turn, have direct or indirect effects on human life. Second, they promote unintended social consequences—consequences that are generally mediated by economic markets (e.g., the replacement of workers by machines or the emergence of boomtowns). Thus common knowledge has it that technologies perform one or perhaps a few intended functions, while also producing a limited range of unintended social and environmental consequences.

Although this view of technology is straightforward, it is also incomplete and misleading. It diverts attention from many significant aspects of technology, including some of central concern to democracy. By synthesizing recent technological criticism, the alternative view of technology introduced here incorporates the accepted view's sound insights but situates these within a broader perspective that recognizes technologies as a species of social structure.

The phrase "social structure" refers to the background features that help define or regulate patterns of human interaction. Familiar examples include laws, dominant political and economic institutions, and systems of cultural belief. Technologies qualify as social structures because they function politically and culturally in a manner comparable to these other, more commonly recognized kinds of social structures. A series of illustrative examples will clarify this notion. (See Figure 2-1 for a summary of some of the terms used to discuss technology.)

TECHNOLOGIES AS SOCIAL STRUCTURES

Chapter 1 introduced Ibieca, the Spanish village that found that its indoor plumbing came at the expense of community integration. That is an instance of a technology helping to structure social relations. Upsetting a traditional pattern of water use compromised important means through which the village had previously perpetuated itself as a self-conscious community.[2] In the United States the automobile has played a somewhat similar role in disrupting prior patterns of community life.[3]

These are not isolated cases; technologies designed for such mundane tasks as commuting to work or cooking food also routinely help constitute social systems of cooperation, isolation, or domination.[4]

> Technology often embodies and expresses political value choices that, in their operations and effects, are binding on individuals and groups, whether such choices have been made in political forums or elsewhere. . . . Technological processes in contemporary society have become the equivalent of a form of law—that is, an authoritative or binding expression of social norms and values from which the individual or a group may have no immediate recourse.[5]

Coercive Compliance

Technologies help regulate social behavior in part because they are themselves governed by both physical and political laws. For example,

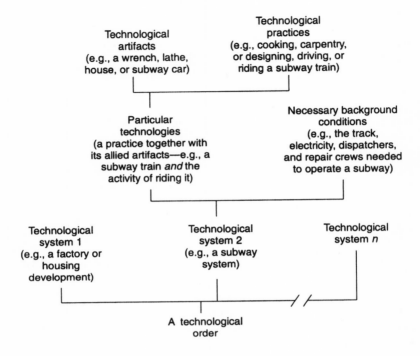

FIGURE 2-1. The hierarchic relationship among basic technological concepts. Additional definition: Technological style = Similarities in function, performance, necessary background conditions, or effects among diverse technologies. n = total number of technological systems within a given community or society.

the operation of many technologies—such as automobiles, medical X-ray machines, or guns—is legally regulated. Thus their misuse can entail a socially enforced penalty.

However, whether or not they are governed by legal regulations, technologies generally embody a variety of other kinds of coercive mandates. The penalty for resisting these mandates may range from an informal reprimand ("Don't lick the food off your knife!") to economic loss or systemic failure (e.g., the gears in a conveyor belt jam, or a worker's hand is injured). These latter results are akin to the consequences befalling those who ignore physical laws (e.g., when someone literally walks on thin ice). Thus physical constraints, or accompanying legal and social sanctions, are among the obvious means through which technologies help structure human behavior.

Subconscious Compliance

Sometimes technologies shape behavior and relationships less through brute compulsion than via subtle, psychological inducement. For example, social scientists have shown that the physical arrangement of chairs and tables strongly influences the kind of social interaction that occurs in schools, nursing homes, and hospitals. Yet the staff in those institutions had previously attributed behavior (including their own) entirely to the mix of personalities and psychological capabilities. They were surprised to learn that simply shifting the furniture could, for instance, help reanimate a seemingly moribund group of mentally impaired hospital inmates.[6]

Opportunities and Constraints

Social structures are also ambiguous in that while they can restrict opportunities in some respects, they can—when appropriately designed—enhance them in others.[7] For example, well-crafted laws help protect basic civil rights and, by providing a relatively stable and well-ordered social context, make it easier for people to realize their life plans.

Besides creating novel opportunities and constraints, technologies also reconfigure prior patterns. For instance, within some offices and factories the proliferation of personal computer networks has enhanced lower level workers' chances to contribute to production decisions while simultaneously challenging midlevel managers' former domains of authority and autonomy.[8] Once deployed, technologies can also aid or hinder the use of other technologies. For instance, telephone systems gradually displaced telegraph services but have more recently facilitated development of computer networks and long-distance data processing.

Background Conditions as Imperatives

In order to function, technologies require various environmental and organizational background conditions. A television set is only useful so long as viewers know how to operate it, it is protected from inclement weather, there is access to electricity, programs are being produced and distributed, and so on.

Frequently when individuals or groups acquire new technologies or technological facilities, they are at best only dimly conscious of the demands that effective operation will impose or require to be developed.

Several years ago a town near mine in western Massachusetts approved construction of an industrial research center, hoping thereby to realize tax benefits. But no one asked beforehand the eventual costs (financial, environmental, and emotional) that the town would one day bear in order to accommodate both new research activities and the concomitant growth in commuting, ancillary employment, and residential population. These costs could include hazards associated with toxic waste disposal, future loss of open space to new housing, and the burden of upgrading roads, sewer lines, snowplowing capabilities, schools, and school bus lines.

To the extent that a given technology plays only a small part in one's life, maintaining the conditions needed for its operation may be of no particular concern. But as a person or society grows dependent on a technology, the necessary conditions of its operation loom as practical imperatives. The need to support these conditions represents a way in which technologies exert a profound structural social influence.

Technology as Structural for Nonusers

Often technologies exert comparably significant effects on people who neither operate nor use the technology in question. One clear example involves phenomena that economists label "spillover effects" or "externalities."[9] Homeowners hear neighbors' radios, lawn mowers, or air conditioners; whole communities breathe noxious fumes from an industrial facility. Each person lives in an aesthetic landscape that reflects the aggregate technological choices made by other people or organizations. The psychological texture of our everyday life reflects the influence of countless technological choices and practices in which we did not participate.

Moreover, often such spillover effects exert a structural influence that is dynamic and transformative. For instance, someone might choose not to purchase a power lawn mower to avoid its noise. However, after a few neighbors have bought theirs, this person may reconsider, thinking, "Since I'm suffering from the noise anyway, why not buy my own power mower and at least benefit from the convenience?" In this way each mower purchased contributes to a cycle that gradually transforms a neighborhood of quiet into one rent by the sound of churning engines.

Next reconsider the background conditions necessary for a technology to operate. Many of those conditions have a tremendous impact on lives even if individuals do not own the technology or use the technological service that establishes their raison d'être. Suppose, to state the case dramatically, that as a citizen of a modern nation a woman opts for a relatively self-sufficient mode of life: She refuses to own a car, uses solar collectors on the roof of her home, and plants a large vegetable garden in her yard. What has she accomplished? Something,[10] certainly, but the texture of her world still reflects the existence not only of cars and their immediate culture, but of roadways, automobile manufacturing and marketing systems, oil refineries, electric generating facilities, agribusiness, the private or public bureaucracies that manage these things, and their often tumultuous politics. That is part of what it means to say that technologies are social structures. The aggregate result of a society's many technological choices in one way or another affects every member.

Communicative and Cultural Systems

Apart from materially influencing social experience, technologies also exert symbolic and other cultural influences. This is true not only of technologies explicitly called communications devices (e.g., cellular phones, televisions, and radios), but of all technologies.

For example, modern sofas generally have two or three separate seat cushions. There is no compelling technical or economic rationale for this design (an affordable, seamless sofa is an easily conceived alternative—as seamless mattresses and Japanese futon sofa-beds attest). Rather, separate sofa cushions define distinct personal spaces and thus respect—but also help to perpetuate—modern Western culture's emphasis on individuality and privacy.[11]

Technologies even play transformative roles within psychological development. For example, earlier this century Swiss psychologist Jean Piaget determined that young children distinguish living from nonliving things according to whether or not the things move, and—as the children develop psychologically—then according to whether things move by themselves or are moved by an outside force. However, more recently, social psychologist Sherry Turkle found that children who play with computer toys that appear to "talk" and "think" develop different criteria for distinguishing "alive" from "not alive." Instead of relying on physical criteria (such as motion), they invent psychological criteria and hypotheses ("Computers are alive because they cheat" or "Computers are not alive because they don't have feelings"). Children's developmental trajectories, including their conceptions of self and moral reasoning, are transformed as a result of their interactions with these machines.[12]

The process that Turkle described with respect to computer toys is a specific instance of a much more general phenomenon. As they

reconfigure opportunities and constraints for action, and function simultaneously as symbols and latent communicative media, technologies also reconfigure opportunities and constraints for psychological
development.[13]

Macropolitics: Technology *and* Society versus Technology *as* Society

Many scholars have described cases in which technologies exert a
macrolevel influence on societies. Consider historians who focus on the
social role of just one or two important technologies at a time. Large-
scale dams and irrigation systems may have played a decisive role in the
creation and maintenance of states in antiquity. Lynn White Jr. told a
now-famous story of the role of the stirrup in the development of
European feudalism: stirrups made possible mounted shock combat,
which led in turn to heavy full-body armor, heraldry, chivalry, stronger
horse breeds, more efficient plowing methods, and so forth.[14] In America, railroads helped establish national markets; promoted coal mining,
steelmaking, and the widespread adoption of steam power; provided an
influential model of geographically dispersed, hierarchically managed
corporate organization; contributed to the adoption of standardized
timekeeping; and served as a dominant metaphor with which Americans interpreted their entire civilization.[15]

More recently in the United States one role of new technologies
has been to provide grounds for the growth of the federal government,
through the proliferation of such agencies as the Federal Communications Commission for regulating telecommunications, the Federal Avia-
tion Administration for regulating the airline industry, the Nuclear
Regulatory Commission and the Department of Energy for administering aspects of national energy production and nuclear weapons development, and the like.

In each of these instances, technological innovation plays a role in
establishing, transforming, or maintaining states or societies at the
macrolevel. Langdon Winner has explored the further hypothesis that
the entire ensemble of modern technological systems—including the
background conditions required to keep them operating—tends to
promote centrally coordinated, technocratic social administration.[16]

Hence there are numerous examples in which technologies affect
societies or states in ways that have macrostructural implications.
However, this formulation—while both true and dramatic—nonetheless misses the force of this chapter's earlier analysis. Technologies
function politically and culturally as social structures by coercing physi-

cal compliance; prompting subconscious compliance; constituting systems of social relations; establishing opportunities and constraints for action and self-realization; promoting the evolution of background conditions; affecting nonusers; shaping communication, psychological development, and culture generally; and constituting much of the world within which lives unfold.

Considering all of the preceding functions and effects together, it would be fairer to say that technologies do not merely *affect* society or states, they also *constitute* a substantial portion of societies and states. That, too, is part of what it means to be a social structure. Recognizing the many respects in which technologies contribute to defining who people are, what they can and cannot do, and how they understand themselves and their world should dispel the common myth that technologies are morally or politically neutral.[17]

Influential, Not Determining

Technologies "structure" social elations in that they shape or help constitute—but do not fully determine—social experience.[18] Water pipes and washing machines did not, for example, literally force Ibiecans to stop gathering at their village's central fountain and washbasin, but instead altered the system of inducements and interdependencies that formerly made such gathering occur naturally.

Aside from the possibility of rejecting or retiring a particular technology, there is always a margin of flexibility in how existing technological artifacts may be used or operated, or in what activities may occur in conjunction with them. This margin is finite, and its extent varies from one technology to the next and over time, but it nevertheless exists. For example, while a conventional assembly line provides only highly restricted opportunities to vary work routines at each station, it does not materially prevent workers from rotating jobs among work stations.[19]

Context-Dependency

Developing a railroad network helped catapult the United States to global economic preeminence, but Britain developed railroads earlier and yet nonetheless gradually lost its world economic predominance. Thus railroads (or other technologies) are socially consequential, but how and why they matter depends on the precise technologies in question in each particular context of use.

Moreover, just as social context—including, among other things, a society's preexisting technological order—regulates each technology's material functions and effects, it also regulates a technology's communicative functions and cultural meanings. A few decades ago a belching smokestack symbolized progress. Today—in a different historical context—the same smokestack is more likely to evoke distress or even outrage.

Finally, one important influence on a technology's functions and effects is the minds and culture of people.[20] Nineteenth-century high-wheeler bicycles were perceived by athletic young men as virile, high-speed devices. But to some women and elderly men the same devices signified personal danger. Indeed, conflicting perceptions of the high-wheeler proved consequential to its subsequent technological development. Its perception as a "macho machine" prompted new bicycle designs with ever higher front wheels. The competing perception of the high-wheeler as an "unsafe machine" prompted designs with smaller front wheels, different seat placement, or higher rear wheels.[21] Thus to understand the social function, meaning, and evolution of the high-wheeler, it is essential to explore its psychological and cultural context.

Public controversies concerning technology offer another occasion for observing the role of culture and cognition in establishing a technology's context, and hence its social role. For example, during the 1970s nuclear engineers and electric utility executives generally viewed centralized production of electricity as a critical social need and essential to the concept of commercial nuclear power. To them an alternative to nuclear power needed to be another means of performing this critical function.[22] But other energy policy analysts saw the expanded production of electricity as so inessential that a perfectly viable alternative could be a panel of foam wall insulation that did not generate any electricity.[23]

In evidence here are fragments of a social process of contesting or negotiating what is or is not to count as an essential function of a technology and hence as an alternative.[24] Thus, when technological consequences or meanings become controversial, processes through which technologies are culturally constituted may emerge openly.

Contingent Social Products

There is residual variability in the structural effects associated with any deployed technology—*within* a particular social context and even more so *among* different contexts. However, a technology's greatest flexibility exists before its final deployment, when artifacts and their accompanying social organization are being conceived and designed.

Technologies do not just appear or happen; they are contingent social products. Thus it is possible, both before and after the fact, to imagine alternative designs. The process by which one set of designs rather than another comes to fruition is influenced by prevailing social structures and forces, including the preexisting technological order. However, this process also reflects explicit or tacit social choices, including political negotiations or struggles.[25]

For example, it is hard to imagine a modern home without an electric refrigerator, but had the accidents of competing corporate resources played out slightly differently, gas-powered refrigerators that would have run more reliably and quietly could have been the norm.[26] Other feasible alternatives in household technology harbored the potential for even more dramatic social effects.[27] Moreover, although today people think of the guiding impulses behind technological development as necessarily being profit, convenience, or military advantage, throughout history religious or aesthetic motivations have often been just as significant.[28]

Thus there are many potential, competing technological pathways, and each is socially developed. But the flexibility associated with a given technology, or with other social structures, tends to diminish with time. After a society has habituated itself to one technology, alternatives tend to become less accessible. Once designed and deployed, a technology, like a law or a political institution tends—if it is going to endure—gradually to become integrated into larger systems of functionally interdependent artifacts and organizations and then to influence the design of subsequent technologies, laws, and institutions such that the latter all tend to depend on the continued existence of the former. Thus, owing to the accompanying evolution of supporting custom, entrenched interest, and various sunk costs, it is often difficult to achieve radical design alterations once an initial decision has been implemented.[29] A further factor reducing the flexibility of technologies is that they exhibit some of the pure physical recalcitrance that comes with material embodiment. Hence, both technologies and other social structures, once they have come into existence, tend to endure. However, technologies exhibit a remaining characteristic that tends to distinguish them from other social structures and to increase their relative political salience: polypotency.

POLYPOTENCY

Technologies function as social structures, but often independently of their (nominally) intended purposes. This is one of the phenomena that the conventional view of technology obscures. The same obfuscation is reflected in studies that profess a broad interest in the political effects of technology but that discuss only technologies designed explicitly to function politically (such as telecommunications, military and police technologies, voting machines, or computer databases).[30] Such technologies indeed function politically, but everyone knows that. That is these technologies' announced purpose. Harder to grasp is the truth that all technologies are associated with manifold latent social effects and meanings, and that it is largely in virtue of these that technologies come to function as social structures. In other words, technologies exhibit superfluous efficacy or "polypotency" in their functions, effects, and meanings. (The word *polypotency*, meaning "potent in many ways," is introduced here for want of a better existing term. The unfamiliarity wears off quickly if one contrasts it with *omnipotence*, meaning, literally, "potent in all ways.")

For example, when a man uses an ordinary hammer to pound nails, he also learns about the texture and structural properties of materials, he exercises and develops his muscles, he improves his hand–eye coordination, and he generates noise, all while stressing and wearing the hammer itself. As his competence at hammering grows, he feels his self-respect affirmed and approved. At another level, his activity resonates with half-conscious memories of primeval myths about Vulcan and Thor. He is also reminded of the blacksmith and the mythology of the American frontier. He thinks of a judge's gavel, the hammer as a symbol of justice, and a song popularized by the folksinging trio Peter, Paul, and Mary.

Where did the hammer come from? Somebody chopped down a tree and fashioned the handle. Others located and extracted iron ore. Some of that ore was refashioned into a hammer head. If a man touches his tongue to the hammer, with the taste of oxidized iron he senses fleetingly a former age when once-independent craftsmen and farmers first found themselves working under strict supervision in a factory. When he was a child, an uncle first taught him to use a hammer. Now when he hefts a hammer, he feels embedded in a historical relationship with this and other hammers and with the development of the concept of hammers and technology in general.

The hammer's immediate social context of use can vary. The man may work alone, on a project with others, or in a room where each person pursues a different project. He may or may not choose his task; he may or may not earn a wage. Depending on the precise social context of its use, the hammer means different things to him, he sees it differently, and it helps disclose the world to him in different ways. Likewise, his style of using the hammer discloses to others much about his character, competence, and mood.

The hammer differs from a partially automated assembly line in that the latter requires and helps coordinate the simultaneous efforts of many workers. But a hammer also establishes certain limiting possibilities on the social conditions of its use. Hammers have only one handle. They are not designed to permit the type of close collaboration that is possible through computer networks or necessary when using a long, two-handled saw.

The material result of the man's activity is likely to include some bent nails, scrap wood, a hearty appetite, maybe a bruised thumb, a few sore but marginally strengthened muscles, some excess exhalation of carbon dioxide, perspiration, and a product that becomes part of the humanly shaped world.

So, is the nail entering the board necessarily the most important feature of the activity called "hammering"? Hammers, like all technologies, are polypotent in their social functions, effects, and meanings.

Today's accepted view of technology takes a step toward acknowledging polypotency by speaking of technologies' unintended or secondary consequences. However, the term "polypotency" is helpful in not presuming that one knows automatically which of a technology's many functions or meanings are the most important or even which are intended. Many social historians of technology have, for example, argued that a latent but intended function of some innovations in manufacturing technology has been to substitute low-paid unskilled workers for higher-paid skilled workers, discipline the remaining workforce, and weaken unions.[31]

It is furthermore useful to introduce the term "focal function" to refer to a technology's (ostensibly) intended purpose. "Nonfocal" then denotes its accompanying complex of additional—but often recessive—functions, effects, and meanings. Thus, 19th-century New England schoolhouses' focal function was to provide a space for educational instruction, whereas one of their nonfocal functions was to help generate—in part via the symbolism of churchlike architecture—a relatively docile workforce.[32] (See Figure 2-2 for a summary of some of this chapter's principal concepts.)

Occasionally technologies function as social structures precisely by virtue of their focal purpose. For instance, weapons function coercively because they are designed to do just that. But more often and more subtly, it is technologies' latent polypotency that accounts for their structural performance. This is illustrated by many previous examples, ranging from sofa cushions (which help to latently reproduce our culture's sense of privacy) to computer toys (which unexpectedly alter children's psychological development). Even technologies focally designed to function structurally are apt to structure nonfocally as well.

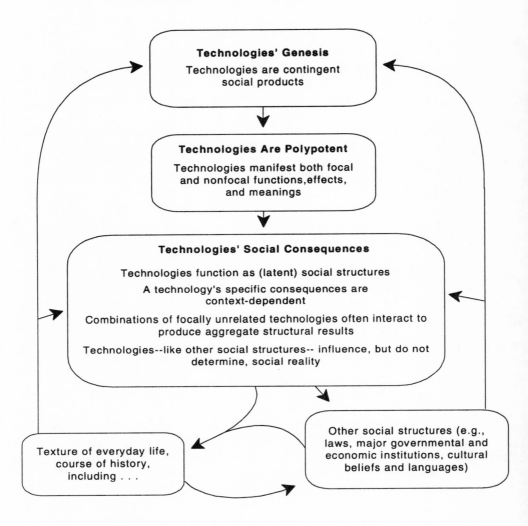

FIGURE 2-2. Technology as a social and political phenomenon.

For instance, nuclear weapons are designed focally to coerce, deter, or destroy other societies, but they contribute nonfocally to legitimating authoritarian government institutions within the societies that possess them.[33] Marshall McLuhan popularized this truth as it applies specifically to technologies focally designated as communications devices: "The medium *is* the message."[34] In other words, the technical means of focally delivering a message can, owing to polypotency, matter more than the message itself.

Moreover, often groups of focally unrelated technologies interact latently to produce a structural effect that no one of them could accomplish alone. Distinct sofa cushions would not help establish cultural norms of privacy and individualism were they not part of a complex of artifacts and ritual behavior that contribute jointly toward that same result. (Other artifacts in the complex with sofa cushions include individual eating utensils, private bedrooms, telephone receivers designed to accommodate one person at a time, and so forth.[35]) In short, to achieve social insight and efficacy, it is essential to consider all the different artifacts and practices that comprise a society's technological order.

There are important functional equivalencies between technologies and nontechnological social structures (e.g., legal statutes, government agencies, and large corporations). All represent enduring social products that shape subsequent social experience. However, there are also differences, revolving around contrasting levels of social understanding with respect to each.

First, laws and political and economic institutions are contingent social products, and at some level everyone knows this truth. Societies evolve these things through formal political or juridical processes, and it is commonly understood that alternative choices are possible. In contrast, people are prone to misperceive a society's technologies as inevitable, that is, as naturally determined rather than socially shaped and chosen.

Second, laws and other formally evolved social structures are commonly understood to function as social structures. That is their explicit purpose. Certainly, they can also be implicated in the production of various unintended social consequences. Prohibition-era laws were enacted to stop alcohol production, not to drive it underground and contribute to the expansion of organized crime. However, people at least expect that legal statutes and institutions will—because that is their intent—in some way shape social interaction and history. In contrast, people ordinarily expect most technologies to prove structurally inconsequential, and—because focally most of them do—this expectation appears confirmed. But here is where appearances deceive, insofar as it is frequently a technology's nonfocal aspects alone that conspire to manifest profound structural consequences.

Hence, although technologies are as consequential as other social structures, people tend to be more blind both to the social origins of technologies and to their social effects. This dual blindness is partly due to certain myths or misconceptions, such as the myth that technologies are autonomous self-contained phenomena and the myth that they are morally neutral.[36] It is also inculcated through modern technologies themselves, via both their style and their social process of design (see Chapter 6).

These dual misperceptions concerning technologies actually enhance their relative structural significance, because they enable technologies to exert their influence with only limited social awareness of how, or even that, they are doing so. This helps explain why people are prone to resign themselves to social circumstances established through technological artifice and practices that they might well reject if the same results were proposed through a formal political process.[37]

So long as their social origin, effects, and dynamics remain so badly misperceived, technologies will not suffer the same liability as would, say, functionally comparable laws or economic institutions, of being challenged on the grounds that they are politically or culturally unacceptable. Furthermore, societies will fail to develop the capacity to seek other technologies more consonant—both focally and nonfocally—with their members' ideals and aspirations.

Chapter 3

"IN EVERY SENSE
THE EXPERTS"
Strong Democracy and Technology

> In West Central Minnesota, local
> farmers have been opposing an
> electrical transmission line for over
> four years. . . . The public relations
> man for the utility said . . . , "You
> should be proud to have the biggest
> powerline in the world in your
> country," but the farmers felt
> differently.
>   To people who love and care for
> the land, a transmission line of this
> size is a desecration. People who
> once felt they lived in a democratic
> society feel they have been betrayed
> and no longer control their own
> lives.
>   —*Minnesota farmer and protester*
>   *Alice Tripp*[1]

How does the preceding chapter's key insight—that technologies represent a species of social structure—bear on the relationship between technology and democracy? The answer depends partly on one's concept of democracy. One common view is that, as a matter of justice, people should be able to influence the basic social circumstances of their lives. This view implies organizing society along relatively egalitarian and participatory lines, a vision that Benjamin Barber has labeled "strong democracy."[2]

Historic examples approaching this ideal include New England town meetings, the confederation of self-governing Swiss villages and cantons, and the English and American tradition of trial by a jury of peers. Strong democracy is apparent also in the methods or aspirations of various social movements such as the late-19th-century American Farmers Alliance, the 1960s U.S. civil rights movement, and the 1980s Polish Solidarity movement.[3] In each of these cases ordinary people claimed the rights and responsibilities of active citizenship concerning basic social issues.

The strong democratic tradition contrasts with more passive or inegalitarian models of democracy that in practice tend to prevail today, so-called thin democracy.[4] Here the focus shifts from a core concern with substantive political equality and with citizens' active engagement in political discourse, or in seeking their common good, to a preoccupation with representative institutions, periodic elections, and competition among conflicting private interests, elites, and power blocs.[5] Within thin democracies power is less evenly distributed; citizens can vote for representatives but ordinarily have little direct influence on important public decisions.

The contest—both in theory and in practice—between the strong and thin democratic traditions is long-standing and unlikely to be resolved soon. Rather than stopping now to compare and contrast the two, I propose initially to suspend judgment and simply posit a specific, strong democratic model of how societies ought to be organized.

TECHNOLOGY AND DEMOCRACY

The strong democratic ideal envisions extensive opportunities for citizens to participate in important decisions that affect them. A decision qualifies as important particularly insofar as it bears on a society's basic organization or structure. The commitment to egalitarian participation does not preclude continued reliance on some representative institutions, but these should be designed to support and incorporate, rather than to replace, participatory processes.

Complementing this procedural standard of strong democracy is a substantive standard: in their political involvements citizens ought, whatever else they do, to grant precedence to respecting any important concerns or interests common to everyone. Above all, they should perpetuate their society's basic character as a strong democracy. Apart from this one substantive moral obligation, citizens are free to attend as they wish to their diverse and perhaps conflicting personal concerns.

This model of democracy, even in schematic form, is sufficient for deriving a prescriptive theory of democracy and technology: *If citizens ought to be empowered to participate in determining their society's basic structure, and technologies are an important species of social structure, it follows that technological design and practice should be democratized.* Strong democracy's complementary procedural and substantive components entail, furthermore, that technological democratization incorporate two corresponding elements. Procedurally, people from all walks of life require expanded opportunities to shape their evolving technological order. And substantively, the resulting technologies should be compatible with citizens' common interests and affinities—to whatever extent such exist—and particularly with their fundamental interest in strong democracy itself.

Democratic Evaluation, Choice, and Governance

The preceding argument suggests that processes of technological development that are today guided by market forces, economic self-interest, distant bureaucracies, or international rivalry should be subordinated to democratic prerogatives. Only in this way can technologies begin actively to support, rather than to coerce or constrict, people's chosen ways of life. For example, residents of many American cities have grown resigned to daily traffic jams, sprawling shopping malls, the stress associated with combining careers with parenthood, and the television as babysitter. This pattern of sociotechnological organization is largely haphazard.[6]

At other times, an existing technological order, or its process of transformation, reflects the direct intentions of powerful organizations or elites. For instance, this chapter's epigraph alludes to an electric utility consortium that proceeded, despite adamant local opposition, to construct a huge transmission line across prime Minnesota farmland. That outcome was not haphazard or unplanned, but neither did it reflect democratic preferences.

Technological evolution can thus encompass social processes ranging from the haphazard to the bitterly contested or blatantly coercive. None of these processes is strongly democratic. This is not to say that every particular technology must suddenly be subjected to formal political review. Each time one is moved to buy a fork or to sell a pencil sharpener, one should not have to defend the decision before a citizens' tribunal or a congressional committee. Not all technologies exert an equal structural influence. However, consider a modern society's treatment of another genus of social structure: various kinds of law. The rules that parents create for their children are subject to relatively little social oversight. But rule making by federal agencies is governed by extensive formal procedure, and even more stringent procedure is required to amend a national constitution. Why should the treatment of technology be so different?

Whether a technology requires political scrutiny and, if so, where and how exhaustively, should correspond roughly to the degree to which it promises, fundamentally or enduringly, to affect social life. This implies the need for a graduated set of democratic procedures for reviewing existing technological arrangements, monitoring emerging ones, and ensuring that the technological order is compatible with informed democratic wishes.

Within such a system, citizens or polities that believe that a set of technologies may embody significant structural potency ought always to have the opportunity to make that case in an appropriate political forum. Beyond this, there should be a system of ongoing democratic oversight of the entire technological order, scanning for the unanticipated emergence of undemocratic technological consequences or dynamics, and prepared when necessary to intervene remedially in the interest of democratic norms.

This does not mean, however, that everyone has to participate in each technological decision that becomes politicized. Logistical nightmares aside, there is more to life than politics. But in contrast to the present state of affairs, there should be abundant opportunity for widespread and effective participation. Ideally, each citizen would at least occasionally exercise that opportunity, particularly on technological matters significant to him or her.[7]

For example, in the early 1970s the cry rang out that there was natural gas beneath the frigid and remote northwest corner of Canada. Eager to deliver the fuel to urban markets, energy companies began planning to build a high-pressure, chilled pipeline across thousands of miles of wilderness, the traditional home of the Inuit (Eskimos) and various Indian tribes. At that point, a Canadian government ministry, anticipating significant environmental and social repercussions, initiated a public inquiry under the supervision of a respected Supreme Court justice, Thomas R. Berger.

The MacKenzie Valley Pipeline Inquiry (also called the Berger Inquiry) began with preliminary hearings open to participation by any Canadian who felt remotely affected by the pipeline proposal. Responding to what they heard, Berger and his staff then developed a novel format to encourage a thorough, open, and accessible inquiry process. One component involved formal, quasi-judicial hearings comprising conventional expert testimony with cross-examination. But Berger also initiated a series of informal "community hearings." Travelling 17,000 miles to 35 remote villages, towns, and settlements, the Berger Inquiry took testimony from nearly 1,000 native witnesses. The familiarity of a local setting and the company of family and neighbors encouraged witness spontaneity and frankness. One native commented: "It's the first time anybody bothered asking us how we felt."[8]

Disadvantaged groups received funding to support travel and other needs related to competent participation. The Canadian Broadcasting Company carried daily radio summaries of both the community and the formal hearings, in English as well as in six native languages. Thus each community was aware of evidence and concerns that had previously been expressed. Moreover, by interspersing formal hearings with travel to concurrent community hearings, Berger made clear his intention to weigh respectfully the testimony of both Ph.D. scientists and teenaged subsistence fishermen.

Berger's final report quoted generously from the full range of witnesses and became a national bestseller. Based on testimony concerning environmental, socioeconomic, cultural, and other issues, the judge recommended a 10-year delay in any decision to build a pipeline through the MacKenzie Valley, as well as a host of more specific steps (including a major new wilderness park and a whale sanctuary). Within months the original pipeline proposal was rejected, and the Canadian Parliament instead approved an alternate route paralleling the existing Alaska Highway.[9]

Some might fault the MacKenzie Valley Inquiry for depending so much on the democratic sensibilities and good faith of one man—Judge Berger—rather than empowering the affected native groups to play a role in formulating the inquiry's conclusions. Nevertheless, the process was vastly more open and egalitarian than is the norm in industrial societies. It contrasts sharply with the steps forced on those Minnesota farmers, mentioned earlier, who were loathe to see a transmission line strung across their fields:

> The farmers have tried to use every legitimate legal and political channel to make known to the utility company, the government and the public their determination to save the land and to maintain safety in their workplaces. The farmers and their urban supporters have been met with indifference and arrogance by both the utility and the government. Turned away at the state capitol, they have taken their case to the courts again and again, only to be rebuffed.[10]

The Berger Inquiry represents just one example of a more democratic means of technological decision making. (For other examples, see Chapter 12.)

Democratic Technologies

Besides fostering democratic procedures for technological decision making, we must seek technological outcomes that are substantively democratic. The purpose of democratic procedures is, most obviously, to help ensure that technologies structurally support popular aspirations, whatever they may be. The alternative is to continue watching aspirations tacitly conform themselves to haphazardly generated technological imperatives or to authoritarian decisions.

However, according to strong democratic theory, citizens and their representatives should grant precedence here to two kinds of aspirations. First and most importantly, technologies should—independent of their diverse focal purposes—structurally support the social and institutional conditions necessary to establish and maintain strong democracy itself. (These conditions are discussed later in this chapter.) Second, technologies should structurally respect any other important concerns common to all citizens.

This does not necessarily mean shifting social resources to the design of technologies that focally support democracy or other common goods. That is the instinct of many strong democrats,[11] and some such efforts may be appropriate. For example, there might be a constructive role within strong democracy for electronically mediated "town meetings."

However, a preoccupation with certain technologies' focal functions, if it excludes commensurate attention to their nonfocal functions and to those of other technologies, is apt to prove disappointing or counterproductive. It might, for example, do little good to televise more political debates without first inquiring whether a nonfocal consequence of watching television is to induce passivity rather than critical engagement.[12] And is it obviously more urgent to seek any new technologies that are focally democratic before contemplating the redesign of existing technologies that, nonfocally, are antidemocratic? How can one know that the adverse effect of the latter is not sufficient to override any beneficial effect intended by the former? For instance, it may be fruitless to try to foster civic engagement via interactive telecommunications unless communities are prepared at the same time to promote convivially designed town and city centers; neighborhood parks, greenhouses, workshops, and daycare centers; technologies compatible with democratically managed workplaces and flexible work schedules; more democratically governed urban technological infrastructures; and other steps toward constituting democratic communities.[13]

In other words, societies do not require a special subset of technologies that are focally democratic as a complement to the remaining majority of technologies that are inconsequential to politics, because the remaining technologies are not, in fact, inconsequential. The overall objective ought to be a technological order that structurally manifests a democratic design style. Considering the entirety of a society's disparate technologies—both their focal and nonfocal aspects—is the technological order strongly democratic? That is the first question.

Owing in part to modern societies' persistent neglect of their structural potency, technologies have never systematically been evaluated from the standpoint of their bearing on democracy. Therefore, upon scrutiny, many existing technologies may prove structurally undemocratic. Furthermore, from a dynamic perspective, they may erect obstacles to efforts intended to further democratization. For example, the declining interest in political participation observed within most industrial democracies might be partly attributable to latent subversion of democracy's necessary conditions by technologies. We can start testing such conjectures after formulating criteria for distinguishing structurally democratic technologies from their less democratic counterparts.

Contestable Democratic Design Criteria

If democratic theory can specify that technologies ought above all to be compatible with strong democracy, does that prescription preempt the most important questions that democratic procedures for technological decision making might otherwise address? No. In the first place, this leaves many important questions to the discretion of democratic judgment. These involve debating shared and personal concerns and then striving to ensure that technologies structurally support them. But even on the prior question of seeking a technological order that structurally supports strong democracy, there is a broad and critical role for democratic involvement.

The simple idea that technologies ought to be compatible with strong democracy is entirely abstract. To become effective, it must be expanded into a sequence of successively more specific guidelines for technological design, what I call democratic design criteria. But to specify such criteria with greater precision and content, and then to use them, one must adduce and interpret a progressively wider selection of evidence and exercise judgment (see Part II). Thus as democratic design criteria become more specific and are applied, the grounds upon which they might reasonably be contested increase.

Moreover, even an expanded system of design criteria will always remain essentially incomplete. For instance, as social circumstances shift or as novel technologies are developed, new criteria will be needed and old ones will have to be reevaluated. In addition, no finite set of criteria can ever fully specify an adequate technological design. Democratic design is ultimately a matter of art and judgment.[14]

This guarantees an ongoing central role for democratic procedure. Democratic theory and its theorists—or anyone else—can help initiate the process of formulating and using design guidelines. However, self-selected actors have neither the knowledge nor the right to make determinative discretionary judgments on behalf of other citizens. Individuals cannot, for example, possibly know what their common interests and preferred democratic institutions are until after they have heard others express their hopes and concerns, and listened to comments on their own, and until everyone has had some chance to reflect on their initial desires and assumptions. Also, individuals cannot trust themselves, pollsters, or scientists to make objective judgments on behalf of others, because invariably each person's, professional's, or group's interests are at stake in the outcome, subtly influencing perception and reasoning. Only democratic forums can supply impartiality born of the balance among multiple perspectives, the opportunity for reflection, and the full range of social knowledge needed to reach legitimate determinations.[15]

Hence it makes sense to seek democratic procedures for formulating and applying rationally contestable design criteria for democratic technologies. These will be "contestable" because the process of generating and refining design criteria cannot be finalized. As technology, social knowledge, and societies and their norms change, one can expect shifts in these design criteria. However, the criteria will be "rationally" or democratically contestable because such shifts need not be arbitrary.[16] They should reflect citizens' current best assessment of the conditions required to realize strong democracy and other shared values. (See Figure 3-1 for the basic ingredients of a strong democratic politics of technology.)

Contrast

The theory of democracy and technology developed here contrasts with predecessor theories that emphasize either broadened participation in decision making or else evolving technologies that support democratic

Process   **Substance**

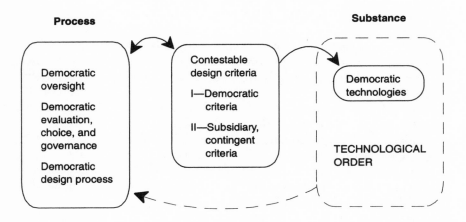

FIGURE 3-1. Democratic politics of technology. A technology is democratic if it has been designed and chosen with democratic participation or oversight and—considering its focal and nonfocal aspects—is structurally compatible with strong democracy and with citizens' other important common concerns. Within a democratic politics of technology, reflection on existing and proposed technologies plays a role in generating democratic design criteria. Use of these criteria then mediates between democratic procedures and the evolution of a substantively democratic technological order.

The figure distinguishes between two categories of design criteria: (I) priority goes to criteria that help ensure that technologies are compatible with democracy's necessary conditions; (II) subsidiary criteria can then reflect technologies' structural bearing on citizens' other concerns and interests. The dashed line indicates that the entire existing technological order exerts a structural influence on politics generally, including (in this instance) the possibility of a democratic politics of technology.

social relations, but that do not integrate these procedural and substantive concerns.[17]

The theory also contrasts with a prevalent view, one that arose during the 19th century, that American mass-production technology was democratic because it made consumer goods widely and cheaply available. Democracy thus became equated with a perceived tendency toward equality of opportunity in economic consumption.[18] This earlier view was insensitive to the structural social consequences associated with production technologies themselves and with the goods and services they produced. Furthermore, as a consequence of this blind spot, the theory foresaw no need to complement the market mechanism for making technological choices with any type of political oversight.

FREEDOM: THE MORAL BASIS OF
STRONG DEMOCRACY

This chapter opened by simply positing a strongly democratic model of democracy; let us now briefly consider a moral argument supporting the model's desirability. Among human goods or values, freedom is widely regarded as preeminent. Freedom is a fundamental precondition of all our willful acts, and hence of pursuing all other goods.[19] But under what conditions is one free? Normally, people consider themselves free when no one is interfering with what they want to do. However, this familiar view is not entirely adequate. Suppose a woman is externally free to pursue her desires, but her desires are purely and directly a product of social conditioning or compulsively self-destructive (e.g., heroin addiction)? How truly free is she?

Such considerations suggest that actions are fully free when guided by something in addition to external incentive, social compulsion, or even a person's own instinctive psychological inclinations. That "something" Immanuel Kant identified as morality—specifically, compliance with moral principles that individuals prescribe to themselves. Morality expresses freedom in ways that cannot otherwise occur, even when one chooses among one's own competing psychological inclinations. With the freedom that morality secures, one acquires the dignity of being autonomously self-governing, an "end unto oneself."[20]

But what should the content of moral self-prescriptions be? Kant envisioned one overarching moral principle, what he called the "categorical imperative." One can think of it as a formal restatement of the Golden Rule: always treat others with the respect that you would wish them to accord you, including your fundamental interest in freedom. In Kant's words, "Act so that you treat humanity, whether in your own person or in that of another, always as an end and never as a means only."[21] Thus, in Kantian philosophy the concept of autonomy connotes moral community and readiness to act on behalf of the common good, rather than radical individualism.

However, suppose I behave morally, but nobody else reciprocates? There would be small freedom for me in that kind of society. Living in interdependent association with others (as people do and must) provides innumerable opportunities that could not otherwise exist, including the opportunity to develop moral autonomy. But it also subjects each person to the consequences of others' actions. Should these consequences seem arbitrary or contrary to their interests, people might well judge their freedom diminished.

As a solution, suppose each person's actions were governed by regulative structures, such as laws and government institutions, that they participated in choosing (hence strong democratic procedure) and that respected any important common concerns, particularly their preeminent interest in freedom (hence strong democratic substance). In a society of this sort, laws and other social structures would each stand, in effect, as explicit expressions of mutual agreement to live in accord with Kant's categorical imperative (i.e., to respect oneself and others as ends).[22]

Strong democracy asks that citizens grant priority to commonalities not for their society's own sake, independent of its individual members, but because it is on balance best for each individual member. Strong democratic procedure expresses and develops individual moral freedom, while its structural results constitute conditions requisite to perpetuating maximum equal freedom. Insofar as it envisions democratic procedures for evolving and governing democratic structures, let us call this a model of "democratic structuration."* In other words, democratic structuration represents strong democracy's basic principle of collective self-organization.

Combine the preceding normative argument with the conventionally slighted insight that technologies function as an important species of social structure. It follows that evolving a democratic technological order is a moral responsibility of the highest order. A democratic politics of technology—one comprising democratic means for cultivating tech-

---

*The basic concept of structuration is that people's thoughts and behavior are invariably shaped by structures that—through ordinary activities (or sometimes extraordinary ones, such as revolution or constitutional convention)—they participate collectively and continuously in generating, reproducing, or transforming. A rough analogy from the natural world can help convey the idea. Consider a river as a process shaped and guided by a structure: its banks. As the river flows, it is continuously modifying its banks, here through erosion, there through deposition of sediment. Over time the river cuts deep gorges, meanders back and forth across broad floodplains, crafts oxbows and bypasses, and establishes at its mouth a complex deltaic formation. Hence the river is a vibrant example of structuration: a process conditioned by enduring structures that it nonetheless helps continuously to reconstitute.

In social life what we do and who we are (or may become) is similarly guided by our society's basic structures: its laws, major political and economic institutions, cultural beliefs, and so on. But our activities nevertheless produce cumulative material and psychological results, not fully determined by structure, that in turn are woven back into our society's evolving structural complex. Hence at every moment we contribute marginally—or, upon occasion, dramatically—to affirming or transforming our society's basic structures.

The word "structuration"—introduced by Giddens (1979, chap. 2)—is not aesthetically pleasing, but it has achieved wider currency than any synonym. I propose the term "democratic structuration" to embed this explanatory concept in a normative context, suggesting that the means and the ends of structuration should be guided by an overarching respect for moral freedom.

nologies that structurally support democracy—is needed to transform technology from an arbitrary, irrational, or undemocratic social force into a substantive constituent of human freedom (see Figure 3-2).

Of course, democracy is by no means the only issue that needs to be considered when making decisions about technology. Citizens might well wish to make technological decisions based partly on practical, economic, cultural, environmental, religious, or other grounds. But among these diverse considerations, priority should go to the question of technologies' bearing on democracy. This is because democracy is fundamental, establishing the necessary background circumstance for us to be able to decide fairly and effectively what other issues to take into account in both our technological and nontechnological decision

I. Philosophical Case for Strong Democracy:

    A. Freedom is a highest order human value. Respecting other people's freedom is a moral duty, necessary for realizing one's own freedom. (Kantian moral theory)

    B. Given the inalterable fact of real-world social interdependence, the opportunity to fully develop and express individual freedom can best be secured within a context of democratic structuration:

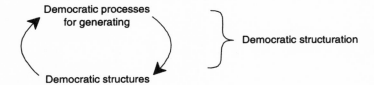

    Under these circumstances both social processes and their structural results, by respecting people's freedom, embody Kantian morality. Here structures support rather than constrain people's highest order interest in freedom. (Neo-Rousseauian, or strong democratic, political theory)

II. Philosophical Case for a Democratic Politics of Technology:

    Applying the preceding argument (I) to technology:

    C. Technologies are a species of social structure. Therefore, it is morally vital that they, like other social structures, be generated and governed via democratic structuration. (The content of democratic structuration, as it applies to technology, is elaborated in Figure 3-1.)

FIGURE 3-2. Philosophical argument for strong democracy and for a democratic politics of technology.

making. (Granting priority to democracy within technological decisions would be somewhat analogous to ensuring compatibility with the U.S. Constitution when drafting or debating proposed laws or regulations.)

It would be presumptuous, however, to insist that the case supporting strong democracy is entirely conclusive. This implies that the contestability attributed earlier to democratic technological design criteria stretches logically back into the supporting theory's philosophical core.

DEMOCRATIC BACKGROUND CONDITIONS

Establishing democratic structuration depends on a number of background conditions. These need to be elaborated here in just enough detail to permit subsequent derivation of democratic design criteria for technologies. The requisite background conditions include: (1) some commonality of purpose, attachment, or outlook among citizens (at a minimum, general recognition of a preeminent interest in living in a strong democracy); (2) some general readiness on the part of citizens to accord higher political priority to advancing important common purposes than to narrower personal concerns; and (3) institutions that foster these circumstances. These background conditions, in turn, incorporate three organizing principles: democratic politics, democratic community, and democratic work (see Figure 3-3).

Democratic Politics

A strong democracy, by definition, affords citizens roughly equal, and maximally extensive, opportunities to guide their society's evolution. What kinds of political institutions does this imply?

Participation and Representation

Is there a middle ground between the present systems of representation, in which a few people participate in deciding most important issues, and obsessive participation, in which *all* people are expected to participate actively in *all* important issues. Barber characterizes the middle ground as a democracy

> in which all of the people govern themselves in at least some public matters at least some of the time. . . . Active citizens govern them-

Democratic politics, democratic community, and democratic work are three organizing principles that ought each, with varying individual intensity, to be present in every basic institutional setting or association within a democratic society:

DEMOCRATIC POLITICS

A. Complementary participatory and representative institutions, within a context of globally aware egalitarian political decentralization and federation (representative institutions designed to support and incorporate direct citizen participation).
B. Respect for essential civil rights and liberties.

To help establish equal respect, collective efficacy, and commonalities:

DEMOCRATIC COMMUNITY

A. Face-to-face human interaction on terms of equality as a means to nurture mutual respect, emotional bonds, and recognition of commonalities among citizens.
B. Intercommunity cultural pluralism.
C. Extensive opportunities for each citizen to hold multiple memberships across a diverse spectrum of communities.

To help develop citizens' moral autonomy, including their capacities to participate effectively in politics and the propensity to grant precedence to important common concerns and interests:

DEMOCRATIC WORK

A. Equal and extensive opportunities to participate in self-actualizing work experiences.
B. Diversified careers, flexible life scheduling, and citizen sabbaticals.

FIGURE 3-3. Some of strong democracy's principal necessary conditions.

selves directly . . . , not necessarily . . . in every instance, but frequently enough and in particular when basic policies are being decided and when significant power is being deployed.[23]

Moreover, on issues in which individuals choose not to participate, they should know that generally others with a similar point of view are participating competently, in effect on their behalf. This may entail, among other things, institutional mechanisms to ensure that the views of socially disadvantaged groups are fully represented and that their needs and rights are respected.[24]

Broadened and equalized opportunities for participation are more than a matter of formal legal rights. They must be supported by relatively equal access to the resources required for efficacy, including time and money.[25] Today, for example, politicians, government functionaries, soldiers, and jurors are paid to perform their civic duties. Why not, when necessary, pay citizens to perform theirs, as did the ancient Athenians? Fairness and equality may also be served by increasing the ratio of representatives chosen by lot to those chosen by vote.[26]

Political Decentralization and Federation

What can help prevent representation from gradually usurping the role of an active citizenry? A partial answer lies in some sort of devolution of centralized political institutions (in which the population's large size renders meaningful participation by all impossible) in favor of a plurality of more autonomous, local political units. By means of small-scale local politics, more voices can be heard and each can carry more weight than in a larger polity. Decisions can be more responsive to individuals, thereby increasing citizens' incentive to participate. There is also potential for small polities to be able to govern themselves somewhat more consensually than can the larger society.[27]

However, various considerations—such as the importance of protecting minorities from local repression—suggest the need to embed decentralization within a larger, federated democratic system.[28] The detailed form of federation must be decided contextually, but its thrust should be toward (1) subsidiarity (i.e., decisions should be made at the lowest political level competent to make them), (2) egalitarianism within and among polities, and (3) global awareness or nonparochialism (i.e., ideally, everyone manifests a measure of knowledge and concern with the entire federated whole—or even beyond it).[29]

In short, power should be relatively diffuse and equal. Political interaction and accountability should be multidirectional—flowing horizontally among local polities, vertically back and forth between local polities and more comprehensive political units, and cross-cut in less formal ways by nonterritorially based groups, voluntary associations, and social movements.[30]

Agenda Setting and Civil Rights

What if citizens are widely empowered to participate in societal choices, but the menu of choice is so restricted that they cannot express their true wishes? Numerous political theorists agree that decision-making *processes* are democratically inadequate, even spurious, unless they are combined with relatively equal and extensive opportunities for citizens, communities, and groups to help shape decision-making *agendas*.[31]

Various civil liberties and protections are also democratically essential either because they are intrinsic to respecting people as moral agents or because we require them in order to function as citizens.[32]

Democratic Community

A strong democracy requires local communities composed of free and equal members that are substantially self-governing. Such communities help constitute the foundation of a decentrally federated democratic polity, and hence of political participation, freedom, and efficacy. They do this in part by establishing a basis for individual empowerment within collectivities that, as such, are much more able than individual citizens to contest the emergence of democratically unaccountable power elsewhere in society (e.g., in neighboring communities, private corporations, nonterritorial interest groups, or higher echelons of federative government). Local communities also provide a key site for coming to know oneself and others fully and contextually as moral agents. The defining features of a democratic community include social structures and practices that nurture collective efficacy, mutual respect, and moral and political equality, and, if possible, help sustain a measure of communitywide commonality.[33]

Strong democratic theory does not envision a perfect societal harmony of interest, sentiment, or perspective. Rather, the central aims of strong democratic practice include seeking amid the fray for any existing areas of commonness; striving to invent creative solutions that, in a just manner, enhance the ratio of concordance to that of conflict; and balancing the search for common purpose against respect for enduring differences and against coercive pressures toward conformity.

Of course, countless forms of community and human association are not locally based. However, strong democracy places a special weight on local community as a foundation (but not a culmination) because of the distinctive and inescapable physical and moral interdependencies that arise at the local level; the territorial grounding of political jurisdictions; and the distinctive quality of mutual understanding, learning, and personal growth that can take place through sustained, contextually situated, face-to-face discourse and interaction.

Cultural Pluralism

If one next considers an entire society's overall pattern of kinds of communities and associations, one discovers that strong democracy does more than *permit* diversity among them: it *requires* diversity. Specifically, democracies should manifest a certain kind of institutional and cultural pluralism: equal respect and protection for all cultures, communities, traditions, and ways of life whose practices can reasonably be construed to affirm equal respect and freedom for all. (Cultures that fail to meet this standard may not warrant unqualified respect, but neither do they warrant determined intervention—unless, that is, they seriously threaten the viability of other, democratic cultures or oppress their own members involuntarily.[34])

There are two principal reasons for this requirement. First, equal respect for people entails respecting their cultural heritage. To undermine a culture corrodes the social bases of its members' sense of self and purpose.[35] Second, all people share an interest in living in a society and a world comprised of many cultures. Cultural pluralism supplies alternative viewpoints from which individuals can learn to see their own culture's strengths and limitations, thereby enriching their lives, understanding, and even survival prospects.[36] Moreover, it provides alternative kinds of communities to which people can travel or move if they become sufficiently dissatisfied with their own.

Democratic Macrocommunity

Democratic politics beyond the level of a single community or group requires generally accepted means of addressing disagreements, and ideally a measure of societywide mutual respect or commonality. The alternatives can include authoritarianism, civil violence, or even genocide—as modern history vividly demonstrates. How, then, can local or association-specific solidarity, together with translocal cultural pluralism, possibly be reconciled with the conditions needed for societywide democracy?

Cultures and groups invariably disagree on fundamental matters sometimes. Nonetheless, there is reason to believe that local democratic communities represent a promising foundation for cultivating societywide respect or commonality.[37] For one thing, ethnic hatred and violence are frequently associated with longstanding political–economic inequalities, not with extant approximations to strong democracy. Moreover, often it is probably harder to escape acknowledging and learning to accommodate differences when engaged in local democracy than in translocal association or politics, where there can be more leeway to evade, deny, or withdraw from differences.

The alternative notion of forging a macropolitical culture at the expense of local democratic communities risks coercion or a mass society, in which people relate abstractly rather than as concrete, multidimensional moral agents. Members of a mass society cannot feel fully respected as whole selves, and furthermore, they are vulnerable to self-deception concerning other citizens' needs and to manipulation by those feigning privileged knowledge of the common good.[38]

One method of nurturing local nonparochialism is to pursue cooperative relations among communities that are distantly located and culturally distinct. There is a good model in those modern U.S. and European cities that have established collaborative relations with communities in other nations regarding matters of peace, international justice, environmental protection, or economic development.[39]

Another route to nonparochialism is to ensure that people have extensive opportunities to experience life in a variety of different kinds of communities. Generally, such opportunities should involve experience both in (1) a culturally diverse array of small face-to-face communities or groups and (2) socially comprehensive, nonterritorially based communities.* The former would encourage concrete understanding of the lives and outlooks of different kinds of people and communities; the latter would provide practical opportunities to generalize and apply what one has learned from these diverse experiences to the problems and well-being of society as a whole.[40]

Nonterritorial associations that are not socially comprehensive—such as ethnic associations, labor unions, churches, single-issue political organizations, and so on—can obviously function as one kind of rewarding community for their members. However, they seem less likely to provoke deep, multiculturally informed comprehension of an entire society.

Democratic Work

People often think of "work" as something they do primarily to earn a living. Here, however, work is interpreted as a lifelong process whose central functions include individual self-development as well as social maintenance (both biological and cultural). "Democratic work" thus denotes (1) work activity through which one can discover, develop, and express one's creative powers, strengthen one's character, and enhance one's self-esteem, efficacy, and moral growth (including readiness to act on behalf of common interests and concerns); (2) a work setting that

---

*The latter are communities or organizations that manifest a multifaceted concern with the well-being of a wide range of kinds of people, if not the entire society or world. Examples include broad-based political parties or movements; federation-level government agencies; and translocal, nongovernmental social service organizations.

permits one to help choose the product, intermediate activities, and social conditions of one's labor, thereby developing political competence within a context of democratic self-governance; and (3) the creation of material or other cultural products that are consistent with democracy's necessary conditions, that are useful or pleasing to oneself or to others, and that thus contribute to social maintenance and mutual and self-respect.[41]

Democratic work contributes richly to individual autonomy and democratic society. Hence, societies cannot be considered strongly democratic if there is involuntary unemployment or if, for example, many people are compelled to work in social environments that are tedious and hierarchically structured, while a few elite managers make important decisions that affect many other citizens.[42] To the extent that good jobs are scarce or societal maintenance requires a certain amount of drudgery or other unpleasant work, vigorous efforts should be made to ensure the sharing of both unpleasant and pleasurable activities.[43]

Diversified Careers and Citizen Sabbaticals

Numerous social thinkers have suggested that people should be able to work in a variety of different careers—either in linear sequence or, preferably, in fluidly alternating succession (sometimes called "flexible life scheduling").[44] However, one reason that is often overlooked concerns the cultivation of citizens' readiness to respect people everywhere as ends in themselves and to act on behalf of societywide interests.

To capture this benefit might require a societywide system, analogous to faculty sabbaticals or the U.S. Peace Corps, that would encourage each person to occasionally take a leave of absence from his or her home community, to live and work for perhaps a month each year or a year each decade in another community, culture, or region. This sabbatical system should include opportunities for the broadest possible number and range of people to take turns within translocal government and administration.

Citizens could then return to their home communities with a deeper appreciation of the diverse needs of other communities, a broader experiential basis from which to conceive of their society's general interest, and lingering emotional attachments to the other communities. (Note, for instance how increased contact between white and African-American soldiers in the U.S. armed services has generally reduced racial prejudice there, thus increasing receptivity to societywide racial integration and equality.[45]) Citizen sabbaticals would thus provide one concrete means of implementing the earlier proposal that citizens have the opportunity for lived experience in a culturally diverse array of communities.

Guiding Principles

Do the three seemingly distinct social domains—formal politics, community, and work—mean that different kinds of basic institutions each contribute to democracy in an essentially different but complementary way? Suppose, instead, one conceives of democratic politics, community, and work as three guiding principles that should each to some extent be active within every basic institutional setting or association (recall Figure 3-3). An actual workplace, for example, may be conceived primarily as a locus of self-actualizing experience and production (work), but it should also ordinarily be governed democratically (politics) and help nurture mutual regard (community).

Failure to embody, within each of a society's many settings and associations, all three principles will tend to result in a whole society much less than the sum of its institutional parts. When, for example, each of a society's basic institutions is merely monoprincipled not only does each fall short of constituting a democratic microcosm, but each in addition tends to stress and overtax the capacities of the others.

Democratic Knowledge

Widespread political participation and the experience of diverse cultures and forms of work amount to an experientially based program of civic education. Living this way, one could hardly help but acquire extensive knowledge of one's world and society. This is not only positive; it is also democratically vital and a civil right. Competent citizenship, moral development, self-esteem, and cultural maintenance all depend on extensive opportunities, available to both individuals and cultural groups, to participate in producing, contesting, disseminating, and critically appropriating social knowledge, norms, and cultural meaning.[46]

Formal politics, in particular, must incorporate procedures that support collective self-education and deliberation. The means might include ensuring multiple independent sources of information with effective representation of minority perspectives; open and diverse means of participatory political communication and deliberation (including subsidies to disadvantaged groups that would otherwise be excluded); and extensive and convenient means of monitoring government performance.

OBJECTIONS AND REPLIES

Let us now consider objections that might immediately be raised against strong democracy.[47] The point is not to imply that the model of strong democracy is unassailable or that there are easy answers to every objection. (Certainly, the preceding background conditions are as subject to rational contestation as are subsequent technological design criteria.) Rather, it is to suggest that it would be a mistake to dismiss the model facilely, especially before seeing whether its further elaboration with respect to technology can help meet some of the objections.

Historically Irrelevant?

Perhaps the most obvious objection is that a vision of strong democracy is so at odds with today's realities as to be historically irrelevant, perhaps dangerously so. One of this objection's premises is quite true. For example, the United States harbors serious discrimination based on differences in race, ethnicity, age, gender, sexual orientation, class, and physical and mental capabilities. The political and economic system grants business corporations a structural political advantage over individuals, communities, and consumer, labor, and civic groups.[48] Bad in themselves, such ills seriously inhibit progress toward achieving greater democracy. For example, long experience of hierarchy in many walks of life establishes adverse psychological and social expectations. Those accustomed to giving orders come to cherish their power and seek more of it. Those used to obeying orders may have trouble conceiving of alternatives, themselves aspire to wielding hierarchic power, resent the powerful rather than oppose the structure of unequal power, or grow resigned to their fate and, with time, suffer impaired self-esteem.[49]

However, do these circumstances undermine a model of strong democracy? Perhaps they provide reason to espouse it more vigorously. Let's consider some potential uses of a strong democratic model under nonideal circumstances:

First, the model furnishes a moral standard or vision from which to evaluate existing societies, including their technological orders. Absent such vision, critique can miss its mark or be less complete. This is reflected in the historic failure to comprehend the breadth of ways technologies affect the democratic prospect. Such impaired awareness, by stymieing social resistance, may well have exacerbated technologies' capacities to structure antidemocratically.

Second, an ideal model can help motivate democratization. Armed with a convincing moral rationale, the weak can begin to bond and grow stronger, while sometimes the strong begin to doubt their own purpose and resolve. So suggests Gandhi's challenge to the British Empire, as well as many hard-won successes by feminists, trade unionists, environmentalists, U.S. civil rights and antiwar activists, and rights' crusaders among the physically and mentally impaired. Morally fortified struggle does not always prevail. But moral vision and conviction are among the greatest resources potentially accessible to the disempowered, the oppressed, and strong democrats generally.[50]

One further objection to the historic relevancy of strong democracy explicitly invokes technology. How often are we told that the ideal of egalitarian participatory democracy has grown passé and that technology (or else the size or complexity of modern societies, which reflect the influence of technology) is centrally implicated in having effected this obsolescence?

> Congress has particular difficulty coming to grips with issues that involve significant input from science and technology. . . . I'm very much afraid it is no longer possible to muddle through. . . . Our democracy must grow up. . . . Jeffersonian democracy cannot work in the year 1980—the world has become too complex.[51]

> A model of the good society as composed of decentralized, economically self-sufficient face-to-face communities functioning as autonomous political entities . . . is wildly utopian. . . . A model of a transformed society must begin from the material structures that are given to us at this time in history, and in the United States those are large-scale industry and urban centers.[52]

Are such commentators correct in judging that modern technology and demographics decisively discredit the strong democratic ideal? Recall that for thinkers such as Rousseau and Jefferson, egalitarian and participatory democracy was a normative ideal—a goal toward which to strive. How, then, can any contingent fact about the world (such as technological complexity) now be understood to refute automatically an ideal vision of how the world ought to be?

Of course, if one discounts the appreciable opportunity that exists for politically informed choice within technological decision making, then facts can logically refute an ideal. That is, if we have no choice about our technologies and their political effects, and our technologies turn out to be incompatible with democratic ideals, then those ideals cannot conceivably be realized. However, once one recognizes that there is substantial social contingency in technological evolution, such conclusions seem less compelling. If a society's technologies are incompatible with an otherwise preferable system of social organization, why not contemplate strategies for acquiring different technologies?

Intrinsically Unworkable?

A second basic objection could be that a model of strong democracy is intrinsically utopian, in the pejorative sense of harboring fatal contradictions or naive fantasies about human nature and institutional competence. Strong democracy does harbor internal tensions. Representation must support and incorporate, but not supplant, participation. The virtues of democratic community must be integrated with a capacity for societywide governance. The search for commonality must not degenerate into enforced conformity, intolerance, or compromises that normalize injustice. Such tensions confirm the need for careful institutional design, vigilance, and ongoing social learning, as well as the necessity for steeling against life's inevitable disappointments and tragedies. But are these tensions more unmanageable than those infecting existing societies or competing models of social order?

How, too, is the strong democratic ideal psychologically or institutionally naive? It does not presuppose that everyone—or for that matter anyone—will become an altruist. Not all social conflict will dissipate. There is no presumption of a decentralized fantasy world in which there would be no administration or need for political coordination among communities, no role for economic markets, and no risk of international strife. In fact, nothing has been assumed about human nature or possible patterns of social interaction that has not already been observed historically.

Finally, the strong democratic model presented here is not overly detailed. Consistent with its commitment to freedom, strong democratic theory should remain relatively abstract, furnishing only a limited number of democratically contestable guidelines and constraints.

Strategically Infeasible?

Is this model of strong democracy too ahistoric, embodying no plausible strategic account to suggest the feasibility of getting from here to there? At this stage in my overall argument, providing such an account would be premature. However, Chapter 12 begins to engage in historically situated strategizing, based in part on the evidence of technology's structural bearing on politics developed in Part II.

Inefficient?

What of the concern that strong democracy could be grossly inefficient—wasteful of time and hazardous to prosperity? As with competing forms of social organization, there are such risks. But this concern may also involve misconstruing the meaning of efficiency and the nature of strong democracy.

An action is efficient if it accomplishes its end without the unnecessary expenditure of scarce resources. However, in a strong democracy social ends are not simply given. They must be formulated via a strong democratic process. Thus, rather than conceivably impairing economic efficiency, democracy is a precondition for legitimately specifying the ends with respect to which efficiency is defined.

But what if democratization would lead to truly severe or catastrophic economic hardship? Both brute survival and basic well-being are conditions of freedom, and thus it would certainly not contravene democratic ideals to retard the pace of democratization under such circumstances. Freedom and democracy can, as highest order social values, legitimately be traded off against economic considerations—but only when the latter both unequivocally and inescapably translate into offsetting threats to freedom and democracy.[53]

Democratic Competence?

The call for extending strong democratic politics into the technological domain immediately elicits further challenges. Aren't most citizens technological illiterates? Isn't it often impossible even for experts to anticipate technologies' social consequences?

The mere fact that citizens will sometimes render flawed judgments is not decisive here because, after all, experts make mistakes too. The normal way of posing the problem of competence is to look at the kinds of questions currently debated publicly about technology—matters of technical feasibility, cost, safety, environmental effects, and so on—and then to ask whether laypeople can be trusted to make reasonable decisions concerning such complicated matters.[54] The standard strong democratic response is yes, especially if people are provided with adequate resources for formulating their own informed opinions. This claim is then buttressed by discussing cases in which lay participants in technical disputes have reached reasonable conclusions.

For example, consider two reports on the acquired immunodeficiency syndrome (AIDS) disease crisis that were released in June 1988. The first, organized by the U.S. National Academy of Sciences, was authored by a prestigious panel of Nobel Prize winners and other scientific experts. In contrast, the second was issued by a group, assembled by the White House to represent "ordinary Americans," that lacked prior expertise on the subject of AIDS. Observers were startled to find that, despite very different initial knowledge bases, the two panels reached similar conclusions across a broad spectrum of issues.[55]

Apart from merely grasping or replicating expert knowledge, laypeople sometimes contribute in their own right to producing new technical information. During Canada's Berger Inquiry, experts' testimony provided essential information concerning technical details of the pipeline proposal and its possible ecological impact. However, there were also crucial gaps in their scientific understanding, and sometimes they lacked the ability to interpret the significance of their own data. It was then that the value of citizens' knowledge garnered through the Bergen Inquiry's community hearings started coming into its own. According to one of the Inquiry's technical advisers,

> Input from nontechnical people played a key role in the Inquiry's deliberations over even the most highly technical and specialized scientific and engineering subjects. . . . [The final report] discusses the biological vulnerability of the Beaufort Sea based not only on the evidence of the highly trained biological experts who testified at the formal hearings but also on the views of the Inuit hunters who spoke at the community hearings. The same is true of seabed ice scour and of oil spills, both complex technical subjects the understanding of which was nonetheless greatly enriched by testimony from people who live in the region.[56]

Finally, those who argue against lay involvement in sociotechnological decision making often seem to be alluding to horrendous decisions and social consequences that they know will ensue.[57] Yet review of actual experience with lay participation does not yield a bumper crop of disastrous decisions.[58] After all, it was not panels of laypeople who designed the Three Mile Island and Chernobyl nuclear plants; who created the conditions culminating in tragedy at Union Carbide's Bhopal, India, pesticide factory; who bear responsibility for the explosion of the U.S. space shuttle *Challenger*; or who enabled the *Exxon Valdez* oil spill.

Thus the standard strong democratic argument—to the effect that laypeople are capable of participating in complex sociotechnical disputes—certainly casts doubt on any facile attempt to disparage the feasibility of a democratic politics of technology. However, it also misses a deeper and more telling point. The argument responds to the contention that the "lay public increasingly confronts a political agenda that . . . only the experts can understand."[59] But is that political agenda, setting aside its accessibility to lay comprehension, the one with which citizens should be most concerned?

The questions that today dominate the agenda of technological politics—the feasibility and cost of accomplishing promised focal results, assessment of ancillary environmental harms, and so forth—are a product of the conventional view of technology. While certainly significant, they overlook many of technologies' latent structural effects and the preeminent concern to ensure that technologies are compatible with democracy.

Naturally, it is reassuring to find instances suggesting that, given a fair chance, lay citizens can grasp complex technical information or contribute their own knowledge to contested technical deliberations. But that is not, in the first instance, what a democratic politics of technology is about. Rather, the first-order question is the structural bearing of technologies on democracy. Beyond that, it is up to democratic processes themselves to formulate further questions rather than to accept anyone's prepackaged agenda.

This turns the tables. If the first order of business concerns democracy and freedom, are technical experts especially qualified to offer answers? Probably not. After all, if today technological decisions are made undemocratically and with inadequate attention to their structural political significance, it is certainly not because scientists, engineers, economists, and other experts grasp the situation, are clamoring to inform the world, but are unable to get a fair hearing. To the contrary, business, government, and the press all routinely solicit the opinions of technical experts, but the latter typically remain oblivious to technologies' politically relevant aspects.

Technical experts are generally preoccupied with physical mechanisms, the principles that contribute to technological artifacts' performance of their focal functions, or the specifically biological or economic consequences of technological operation. But these are not normally the aspects of technology most salient to the possibility of democracy (see Part II). Moreover, as noted earlier, knowledge concerning technologies' democratic and other structural implications is ineradicably value-laden and specific to particular cultural contexts, and thus should never be a candidate for monopolized production by supposedly impartial experts.[60]

Finally, because technical experts enjoy a privileged position within today's inegalitarian political and economic structures, they tend to share with other elites an unstated, and usually quite unconscious, interest in suppressing general awareness of technologies' public, structural face. If that awareness were to surface publicly, the obvious structural salience and value-ladenness of technological decision making would tend to erode a principal basis on which experts have traditionally been accorded a measure of political deference. This interest in suppression arises despite the fact that many participating experts are behaving consciously with complete sincerity and the best of intentions.

The truth that technical experts are quite *inexpert* on the most important social questions concerning technology should effectively challenge restrictions on lay participation grounded in the claim that the experts know best. The point is not that experts ought never to occupy a distinctive niche within technological politics. However, if they do, that special niche should be decided by means of strong democratic procedure, with due sensitivity to experts' deep and abiding shortcomings on the specific subject of democracy and technology, and in a manner that ensures that, with respect to experts, lay citizens reclaim their rightful political sovereignty, formally and in practice.[61]

If experts are ill-equipped for perceiving the dimensions of technology that bear on the practicability of democracy, many nonexperts do not suffer the same liability, or at least not in the same way or to the same extent. First, everyday citizens have a greater objective interest than do privileged experts in advancing strong democracy. It is their own freedom, self-esteem, and well-being that are at stake (whereas, under strong democracy, some undeserved expert privileges might be at risk). Hence laypeople, once democratically assembled, are more likely to be democratically sensitized and to stand on the lookout for developments that could structurally affect their own empowerment.[62]

Second, nonexperts are apt to be aware from their everyday technological involvements of at least some of technologies' broader social conditions and effects, because the latter directly affect their own life experiences. Upon reflection people know, for example, which technologies require or enable them to work with other people and which have to be used in relative isolation; which are challenging or fun and which induce boredom or apathy; which increase local autonomy and which render localities dependent on faraway organizations. But such intuitions receive scant acknowledgment or affirmation and no interpretation of their structural political significance within the discourse that today dominates technological decision making.

Of course, experts are citizens too, and as such may share some of the experiences and sensibilities of nonexpert citizens. However, here structural distortions tend to interfere. Even if they are privately aware of some of a technology's broader social or political effects, technical experts have an interest in arguing that the most important questions at issue are precisely those in which they themselves are expert. For instance, expert participants in public controversy concerning genetic engineering have tended to reduce the entire issue to a matter of assessing specific biological risks, whereas lay participants have drawn attention to accompanying ecological, social, and ethical concerns.[63]

In addition, experts' similar social backgrounds, self-selection in career choices, professional socialization, and tendency to acquire specialized competence at the expense of integrative knowledge and experience, render them statistically unrepresentative of the values and outlook of the larger citizenry.[64] Moreover, technical experts—especially the most influential, elite experts—tend to live experientially far removed from the everyday events and concerns of nonexpert, nonelite citizens. Elite experts are not asked to perform boring, repetitive tasks at work; few have to contend daily with inadequate mass transit systems, inferior housing, or poor medical care; they may even grow out of touch with the reality that, unlike themselves, other citizens do not have direct, easy access to the highest echelons of business and government decision making. Their world is (for them) accommodating, affluent, and already strongly democratic. For these reasons experts—even polarized, contending groups of experts—cannot be counted on to understand or communicate the everyday knowledge and judgment of other citizens.

Third, an important component of any democratic politics of technology involves identifying commonalities among citizens and the impact of technologies on them. Here there can be no question of bypassing broad citizen involvement. To establish their differences and commonalities, citizens must communicate with one another. To discover a technology's social functions, effects, and meanings, one must consider the technology in its context of deployment. But affected citizens are part of that context and know things about it that no one else can. Once again, the experience of the Berger Inquiry is pertinent:

> When discussion turned to . . . complex socioeconomic issues of social and cultural impact, [native] land claims, and local business involvement—it became apparent that the people who live their lives with the issues are in every sense the experts. . . . [At] the community hearings, people spoke spontaneously and at length of their traditional and current use of the land and its resources. Their testimony was often detailed and personal, illustrated with tragic and humorous anecdotes.
>
> To the experts' discussion of problems and solutions, local residents were able to add comprehensive and vivid descriptions of the meaning of an issue in their daily lives. Their perceptions

provided precisely the kind of information necessary to make an impact assessment.[65]

Where expert testimony tended, moreover, to treat the proposed MacKenzie Valley pipeline as an isolated occurrence, natives and other nonexpert citizens insisted on treating it as a probable trigger for subsequent industrial development. Thus, they, rather than the experts, better articulated the phenomenon of systematicity—the tendency for technologies to elicit the emergence of necessary background conditions and other cascading effects.[66]

It is ironic. Today leaders among our technical elite, such as former Massachusetts Institute of Technology president Paul Gray, argue that scientific and technological illiteracy have reached epidemic proportions, threatening national economic well-being and democracy itself.[67] According to the Clinton administration, "The lifelong responsibilities of citizenship increasingly rely on scientific and technological literacy for informed choices."[68] However, if the most important knowledge about a technology involves not its internal principles of operation but its structural bearing on democracy, then presumably the latter kind of knowledge should constitute the very core of technological literacy. Yet experts, even the elite, typically know little about this first-order issue— not even that it *is* an issue. Must one not reluctantly include among the technologically illiterate—in that term's socially most meaningful sense—the majority of technical experts?

A democratic politics of technology will make demands on individuals' knowledge and judgment. It requires institutions through which people can educate themselves and learn to deliberate together. The Berger Inquiry, for instance, devised community hearings and provided funding to disadvantaged participants. Thus reason and evidence alike suggest that lay citizens can rise to the occasion.[69]

Unpredictable Consequences?

One sometimes encounters the further objection that no one—neither laypeople nor experts—can foresee technologies' social consequences, and hence the ambition to develop democratic technologies and decision making is naively impractical. One part of this concern—namely, that it can be difficult to make predictions—is well founded.[70] Predicting accurately is rarely easy and sometimes impossible. But the consequences of this fact for a democratic politics of technology are not so obvious.

First, even if prediction were utterly impossible, this would not be a telling argument against technological democratization specifically. Those objecting usually assume that in expressing their misgivings, they are defending the status quo against the dangers of misguided democratic idealism. But if one cannot anticipate technologies' social consequences, then why be complacent either about existing, relatively nondemocratic, decision-making processes or about permitting technological innovation at all? Thus, if the doubters' concern stood as a serious challenge to the vision of technological democratization, it would equally challenge the status quo.

Second, sociotechnological predictions are often unreliable, but not invariably so. Some reasonable predictions were, for example, made at an early stage concerning the telephone, the airplane, and atomic energy.[71] Those who completely discount the feasibility of prediction often seem to be making several dubious assumptions. One is to suppose that experts and decision makers have for decades been struggling to anticipate technologies' democratic consequences but have found the task infeasible. That assumption is false. Relatively few experts and decision makers are even sensitive to technologies' structural dimensions, and no one anywhere has ever made a systematic effort to anticipate alternative technologies' structural bearing on democracy.[72] It may prove virtually impossible to foresee technologies' democratic consequences, but that is not yet known, because no one has made the attempt.

Also, skeptics often seem to be reasoning by analogy with contemporary experience in predicting conventionally highlighted technological impacts, such as risks to the environment or human health. In such areas one typically confronts the need to estimate the probability of extremely rare occurrences, to assess the biological consequences of exposure to minute levels of potential toxins, or to understand the behavior of extremely complex ecosystems. Information and consensus are notoriously elusive.[73]

Decision making concerning technologies' democratic implications may sometimes involve questions of comparable complexity, but often it will not. Thus if it is difficult to anticipate the toxicity of a new chemical or the environmental effects of carbon dioxide buildup in the atmosphere, it is not necessarily so hard to predict whether a new work process will involve drudgery or whether or not a new economic activity will contribute to local self-reliance. Yet the latter kinds of questions are not only potentially more tractable, but they also better typify the first-order agenda for a democratic politics of technology (see the design criteria developed in Part II).

Furthermore, it is at least conceivable that our ability to predict will gradually be enhanced, although never perfected, through better understanding of technologies' social dimensions.[74] It may also improve simply by our learning to ask better questions. Publicly debating democratic design criteria offers a promising means for learning collectively to perceive technologies' nonfocal aspects and to formulate democratically essential questions.

Third, and most importantly, it is a mistake to assume that a successful democratic politics of technology depends heavily on long-range prediction. After several centuries of rapid technological innovation, our world has become technology-saturated, but it has not been carefully assessed or governed using democratic norms and structural criteria. Thus one of society's deepest needs is not to predict future consequences, but instead to observe and evaluate the technologies we already have. This might lead to promoting, reforming, or replacing certain existing technologies and systems rather than simply integrating new technologies into a taken-for-granted, preexisting technological order.

The world, moreover, is not culturally or technologically homogeneous; indeed, one can learn much about technologies' social consequences not by prediction, but by studying and comparing technologies as they exist, or formerly existed, in diverse sociocultural settings.[75] The concept of nonfocal structural consequences highlights the potential utility of such comparisons, for it suggests that—relative to headline-grabbing "cutting-edge" technologies—more prevalent, mundane technologies are not so socially inconsequential as we have come to suppose. In other words, the coming wave of biotechnologies, nanotechnologies, cryogenics, and information superhighways matters, but so do more familiar technologies such as pipes, washing machines, air conditioners, jackhammers, and electricity distribution networks.

To the extent that it does, however, become important to understand a particular new technology's social implications, one option would be to run voluntary social experiments. By observing technologies on a trial basis in selected communities, it should be possible to discover social consequences that would otherwise be hard to anticipate. There are already partial precedents for such experiments. For instance, many businesses test-market new products and systems prior to full-scale deployment, and pharmaceutical companies must run rigorous clinical trials before marketing new drugs. Democratic trials would be analogous but would also differ in some respects. They would require a more egalitarian design and evaluation process, and their purpose would be to evaluate technologies' democratic and other broad social consequences, rather than only economic marketability or medical efficacy and safety. (Trials are promising for detecting some social consequences but not those that are scale-specific. For instance, local trial of a prototype invention will not disclose impacts that might later arise from creating large production facilities, supporting technological infrastructures, or a new government bureaucracy to regulate use of the proliferating devices.)

One rough indication of the practicability of acting on information garnered from sociotechnological trials involves historical instances in which elites or powerful organizations stifled innovations that experience showed to be inimical to their interests. For instance, Japan's 17th-century warrior aristocracy suppressed the use and manufacture of guns. Among other things, sword-bearing samurai objected to being shot by peasants.[76] Thus potent actors seem able, when self-interest so dictates, to discriminate actively among competing technologies. A more democratic society would differ in striving to empower all members and groups to participate in such technological discernment.

Finally, one democratic design criterion developed in Part II seeks technological flexibility, so that if future technologies should manifest unexpectedly adverse structural effects, they can be modified or replaced without undue hardship.

CONCLUSION

The basic practicability of a democratic politics of technology seems well demonstrated by the accomplishments of the Old Order Amish. As Chapter 1 explained, the Amish have for centuries guided their social and technological development with unusual self-awareness, while retaining their communities' self-governing, democratic character.

Their success suggests that the Amish have learned better than many other societies to perceive technologies' nonfocal, structural dimensions. For instance, their reasons for favoring horses over tractors include recognition that horses reproduce themselves; produce manure fertilizer; compact the soil less; do not mire down in mud (permitting earlier springtime plowing); and help avert dependence on petroleum products, parts suppliers, and outside mechanics. Also, using tractors would lower farm employment (which the Amish wish to maximize) and reduce the overall need for horses, thereby increasing the chance that cars would gradually replace horses for nonfarm transportation.[77]

The Amish style of technological decision making exemplifies a number of the strategies proposed above for coping with the difficulty of prediction. For instance, the Amish observe the social effects of technological innovations in surrounding non-Amish communities and then use those observations as one basis for making their own decisions.

Generally, community members are permitted to adopt a new technology unless that innovation has already specifically been prohibited. However, sometimes the Amish put new technologies "on probation," allowing only tentative adoption so that the social results can be discovered. For example, in various Amish communities once-probationary private telephones, electric generators, and computers were eventually forbidden. On the other hand, Amish communities in east central Illinois ran a one-year trial with diesel-powered bulk milk tanks before deciding not to prohibit them.[78]

If nonexperts are incompetent to participate in technological decision making, if it is impossible to learn to perceive technologies' nonfocal aspects, and if the difficulty of prediction poses an insurmountable barrier to the practicability of a democratic politics of technology, then somebody had better go tell the Old Order Amish.

PART II

Design Criteria for Democratic Technologies

Regarding technology, a basic concern of the Old Order Amish is: "How can we best maintain our Amishness?" Although admirably sensitive to technologies' structural influence, this question is obviously culture-specific. Other people's tasks are not to become Amish but to learn to establish technological conditions that will support their own collective capacity to decide the kind of society in which they want to live. This entails striving toward a societally appropriate, democratic politics of technology: (1) democratic procedures for evolving (2) democratic technologies. Part II tackles the latter component, initiating the process of developing a set of contestable criteria for identifying democratic technologies.

The first two criteria are concerned, respectively, with democratic community and democratic work. The remaining criteria consider various ways in which technologies can influence the establishment of democratic politics. The eventual goal is an evolving set of criteria capable, when democratically deployed and acted upon, of helping to generate democratic technological orders.

Chapter 4

"ACTIVELY RELATED TO THE WHOLE WORLD"
Technology and Democratic Community

Chapter 3 suggests that it is morally vital for human relations to reflect equal respect for each person as an end. Strong democracy requires, moreover, that communities and organizations strive to achieve self-governance and political equality, serving as mediating structures that help empower individuals within the broader society. Finally, freedom can be realized more completely to the extent that citizens share some commonalities. What does technology have to do with these basic requirements?

Chapter 2 argues that technologies are contingent social products that influence and help constitute systems of social relations. This technological patterning can, for example, be a result of the constraints and opportunities that technologies present, of subconscious influences, or of technologically expressed symbolic meanings.

A basic technological design criterion follows. Insofar as (1) social relations should reflect moral and political equality and help nurture commonalities, and because (2) technologies are contingent social products that influence and help constitute systems of social relations, it is (3) democratically vital that technologies nurture equal respect, political equality, and—to the extent possible—sources of social commonality.

This derivation is abstract. Let us therefore recast the criterion in terms of a typology of basic forms of technologically mediated social relations: authoritarian, individualized, communitarian/cooperative, mass, and transcommunity technologies.

Authoritarian technologies are those that help to establish or maintain illegitimately hierarchical social relationships. *Individualized technologies* are used under conditions and for purposes that individuals have substantial liberty to determine for themselves. *Communitarian/cooperative technologies* help to establish or maintain egalitarian, convivial, or legitimately hierarchical social relationships.[1] (I propose that, in general, hierarchies are democratically legitimate if they support a more free and equal society than could otherwise exist. For instance, hierarchy would seem morally justified if there is institutional assurance that [1] those empowered within a hierarchy are using their power in the preeminent interest of advancing subordinates' moral development, while not themselves becoming morally corrupted by their exercise of power; and [2] a more egalitarian social structure would reduce subordinates' opportunities to develop morally.[2])

Mass technologies combine elements of the preceding three: a relatively small, powerful elite is distinguished from a large mass (the authoritarian element). There is, however, rough equality among members of the mass (a limited communitarian/cooperative element). But this equality is relatively thin and insubstantial insofar as direct interaction among the mass tends to be nonexistent or else is mediated through technologies that the elite control (hence, a certain form of individualism). Thus for members of the mass, social relationships tend to be atomized, standardized, or latently oppressive. Finally, *transcommunity technologies* help establish or facilitate democratic social relations among communities or among the members of different communities.

Using this typology to reformulate the earlier, abstract design recommendation, I suggest:

> Democratic societies should seek a balanced mixture of communitarian/cooperative, individualized, and transcommunity technologies, while avoiding technologies that establish authoritarian social relations. Deploying a limited number of democratically governed mass technologies may also be acceptable.

This is Design Criterion A. For clarification and to help establish its practicability, consider a sequence of case studies.

CONTEMPORARY TECHNOLOGICAL ORDERS

In modern industrial societies, most human relationships fall into one of several types: authoritarian, extensively individualized, or else a combination of the two (including mass variants). These relationships tend to be mirrored in the architecture and technologies with which they have historically coevolved.[3] For instance, at school teachers have traditionally seated themselves at large desks up front of, not among, the students. Students' much smaller desks are typically arranged in rows facing the teacher's desk, thereby reinforcing the "natural" expectation that the teacher should be the focus and mediator of all classroom interaction.

Similarly, the physical layout and technologies of workplaces often reflect antagonistic, hierarchical relations of production. In many modern office buildings, for example, outer rooms with windows belong to upper-level employees. The executive head has the largest office, often in a corner space with two sets of windows. Secretaries frequently have no privacy at all and sit in a space with no windows. A former insurance company employee gave this account of the social functions of his office's physical layout:

> The Newark office in which I worked was a typical medium-sized insurance company office. My desk was in the middle of a large open floor of an office building. . . . Most of the windows to the outside world were in the offices of the managers. Those offices were arranged around the outside of the floor. Not only could these managers look outward, but they could look inward as well: each of them had windows overlooking the open floor. The desks on the open floor were arranged in columns and rows. . . . Here 150 employees did their daily work.
>
> Around the matrix was a series of desks turned sideways to the matrix. These were the desks of supervisors or of the secretaries to the managers in the windowed offices. . . . Employees are overseen by supervisors; supervisors are in turn overseen by an office manager in his glasswalled office. . . . The physical arrangement of the workplace is a primary means of control. Managers can, simply by raising their heads, oversee the activities of workers on the floor.[4]

In this case the physical design of the office made social isolation and hierarchical control of the workers possible but not necessary. Under altered social circumstances it would have been easy for workers at adjacent desks to converse, alter the arrangement of desks, or erect partitions. Contriving an outside view for all might, however, have proven difficult.

In many workplaces, direct visual monitoring is now being supplanted by computerized monitoring, for each worker is supplied with a keyboard connected to a centralized computer or network.[5] Of computerization in his workplace, the manager at a large telecommunications company commented:

There are fewer arguments now, hardly any at all. There is less friendship, too. We were once a tight little group. Now we are mechanized. It is impersonal now. You are no longer close to the people you work closely with.[6]

However, as at the Newark insurance company, it is also technically possible—if not managerially probable—to implement computerization in ways less damaging to sociability.[7]

As the preceding examples suggest, from youth onward work often involves a continuous immersion in social settings that induce unconscious accommodation to patterns of dominance and an uncritical acceptance of the apparent inevitability of technological forms that reinforce those social patterns.[8] Authoritarian technologies establish unequal status and respect among people (contrary to the categorical imperative) and tend to block the formation of convivial relations (as authorities, wittingly or not, use organizational and technical means to help undermine intraorganizational solidarity among subordinates).

For women who work primarily as housewives, the story is somewhat more complex. Consider middle-class women in the United States. At school boys and girls have traditionally undergone a similar indoctrination into accepting hierarchy but, during the middle decades of the 20th century, middle-class housewives' work tended to evolve from authoritarianism to individualism. In 19th-century America, a middle-class wife was typically the manager of several lower-class domestic servants, who were often recent immigrants. As the economy industrialized, women who would formerly have been servants were drawn first into manufacturing as unskilled and semiskilled laborers and later into offices as secretaries and clerks. These competing career opportunities increased the cost of hiring domestic labor, tending to deprive middle-class wives of their former help and of their status as lords within a minifiefdom. The gradual result was that the factories in which lower-class and lower-middle-class women now worked began to produce "labor-saving" home appliances, which made it possible for one middle-class wife to do the work that had previously been performed by the several servants whom she had once overseen.[9]

The work performed by housewives has not infrequently been more varied and intrinsically creative than some of the unchallenging labor performed by both men and women in factories and offices. But because it was culturally stigmatized as low-status work, it tended to leave housewives isolated from one another and from equal participation in the formal wage-labor economy.[10] Some of housewives' resulting feelings of isolation have, in turn, been assuaged technologically—for example, through the televised depiction of the less isolated (and much more exciting) lives of soap opera characters.

More recently, the women's movement has partially broken down rigid sex-role stereotyping within the productive division of labor. Women can now compete (but often not on an equal basis) for jobs formerly performed only by men, while some men now play a larger role in household labor. Nonetheless, many women find that microwave ovens, automatic coffee makers, and the like succeed in making it just barely possible to maintain a full-time paying job, while continuing to bear primary responsibility for unpaid housework and childrearing.[11]

Today much of nonwork life tends to be spent in private cars or in private islands of pseudo-self-reliance (homes); shopping in antiseptic and anesthetizing supermarkets, department stores, eating factories (e.g., McDonald's), and malls; and passively observing mass spectator events (sports, movies, and television). As Chapter 2 noted, contemporary individualism is expressed and reproduced via a dense system of artifacts and allied practices, ranging from sofas with distinct cushions to personal computers, automated bank-teller machines, cellular telephones, portable compact disk players, and home air conditioning and central heating. (Household air conditioning keeps people isolated indoors during hot months; central heating and the proliferation of household television sets disperse individual family members who formerly gathered around a common hearth or stove during cold months.) Meanwhile, mass-oriented activities are, as I suggested earlier, appreciably authoritarian. They afford little experience of democratic community or of creative participation in seeking or working for a common good.[12]

Even the 1970s North American "appropriate technology" movement did little seriously to challenge the organization of life into individualized and authoritarian sociotechnological environments.[13] Publications such as *The Mother Earth News, The Whole Earth Catalog, Five Acres and Independence, Other Homes and Garbage,* and *New Roots* projected an idealized image of voluntary withdrawal from city and factory to a self-contained home in the woods that few people could afford to build. But the movement offered little discussion of the realities of concentrated economic power and little concern with political engagement, fairness, and cooperative activity—not to mention any effort to establish federated communities of free and equal citizens.[14]

Of course, this is an exaggeration. There are emerging examples of firms organized with flattened hierarchies and less distinction between managerial and worker responsibilities.[15] Moreover, within advanced industrial societies some people do sing and play music in groups, tell stories around the campfire, belong to consumer cooperatives, contribute to religious-based social services, work in worker-owned and self-managed businesses, and play team sports. But such activities are marginal or nonexistent in the lives of most people.

Within modern societies the relationship among authoritarian work lives, individualized home lives, and mass-consumer-spectator lives appears reciprocally interdependent. In societies reliant on competitive market economies, such an organization of life tends to self-reproduce and is also functional for maintaining capitalist market relations and hierarchical social relations generally. From the viewpoint of businesses, production for either a mass market or for highly differentiated individualized markets eliminates undesirable economies of scale in consumption. For instance, six nuclear families with six high-tech kitchens adds up to a lot more appliance sales than would one common kitchen for the same six families.[16] Similarly, early television sets may have contributed to local community-building when friends and family assembled in the evening around the first set to arrive in their neighborhood. However, with the subsequent mass proliferation of television sets—first, with one to each household, and more recently with one per household member—even gatherings among members of a single nuclear family can become rare.

Both individualized and authoritarian technologies are politically functional for maintaining authoritarian political and economic relations because they limit opportunities for establishing community-oriented political life. Thus there is little institutional context for critical democratic discourse or for collective action that might seek to challenge and democratize sociotechnological systems.

This general disregard for egalitarian community life is also functional in preserving a mobile labor force: people not emotionally tied to a particular community are more willing to move to find a new job or to accept being uprooted when transferred. That mobility, in turn, makes it difficult to maintain stable communities. In the typical American suburb, "there are no sad farewells at the local taverns or the corner store because there are no local taverns or corner stores."[17]

The systemic complementarity between authoritarian and individualized technologies seems inadequately appreciated by critical historians of technology (e.g., Lewis Mumford), most appropriate technology enthusiasts, and social theorists who foresee a kind of natural evolution toward a more convivial economic order.[18] The complementary individualizing and authoritarian forces in industrial societies make it difficult to build and sustain cooperative/communitarian enterprises—which are needed to help ground the construction of democratic communities—on an ad hoc basis, and thus without concerted political effort. But is a more democratic alternative conceivable?

COMMUNITARIAN/COOPERATIVE ALTERNATIVES

That relationships of conviviality could conceivably be an ordinary part of life becomes clear when reading ethnographers' descriptions of so-called primitive societies. To sketch the possibility of a technological style more compatible with the ideals of equality and democratic community, let us first examine communitarian technological practices within a nonindustrial society, and then look at some examples of communitarian/cooperativism more closely approximating modern expectations regarding material comfort, diversity, and privacy. Because concrete examples invariably exhibit local idiosyncrasies, the goal here is to abstract general insights into the possibility of communitarian/cooperative technologies in a wide variety of contexts.

The Amazonian Mundurucú Indians provide one model of a community-oriented technological style that contrasts markedly with the authoritarian and individualized styles that typify industrial societies. The description that follows is based on fieldwork undertaken by anthropologist Robert Murphy. When Murphy knew them, some Mundurucú had largely assimilated into Brazilian culture, but others continued to live in their own villages of 25 to 90 people.

These small villages consisted of a circular arrangement of two to five dwelling houses, a large "men's house," and a shed used by the village women for flour preparation. Mundurucú life was characterized by extensive cooperation and collectivism. The Mundurucú's preferred hunting technique—surround, chase, and ambush—relied on group effort. Group fishing, using natural vegetative poison in a cooperative effort between upstream (poisoner) and downstream (fish collector) groups, was also important. Whichever man supplied the poison became that day's group coordinator. Murphy and his wife participated in one fishing expedition that involved more than 100 Indians from four villages who caught 2 tons of fish.

Horticulture was another important food source. Each married man had one or two garden plots. Because the Amazonian forest soil is poor, new plots had to be cleared each year or two. All the men of a village, working together, did the heavy clearing. The man whose plot was to be cleared acted as the supervisor, although beyond choosing plot location and size, little supervision was required.

The participants generally dawdle on the way to their work, bathing, gathering fruits, and chatting. The actual labor is frequently interrupted by joking, conversation, eating and little jaunts into the woods in search of parrots and pigeons. . . . Ethnographic literature presents many similar instances of collective labor for purposes of enjoyment and cooperation.[19]

The several women of each dwelling house, working collectively, did most of the planting. However, all of the members of a village, working in a single group, planted manioc, the principal dietary staple. Women working together harvested and then processed the crops in a single communal work shed. Work among the women was cooperative and convivial in much the same manner as that among the men:

Mutual interdependence in *manioc* processing is a strong integrating factor among the women. But the *farinha* [manioc-flour] shed offers other gratifications than the purely economic. The women talk and joke as they work there, and those not immediately engaged in labor sit about and join the conversation.[20]

Overall, Mundurucú fellowship was expressed and supported in numerous ways. For example, cooperation was reinforced by a belief that lone people are vulnerable to attack by evil spirits. Mundurucú solidarity and empowerment vis-à-vis the outside world were, however, especially maintained through their communitarian/cooperative technological style:

Mundurucú communities have been persisting despite more than a century and a half of contact with Brazilian society and a constant population drain through emigration and epidemic disease. Some of the principal bases of this cohesion lie in the social means by which Mundurucú technology is implemented, for these provide fundamental ties of personal and familial interdependence.[21]

In contrast with some other traditional societies that display relatively complete equality among the sexes,[22] Mundurucú society exhibited a strongly developed sexual division of labor. (Although ideologically subordinate to men, Mundurucú women were, in fact, through their substantial control of key subsistence activities, more powerful in many ways than the men.[23]) In other respects, though, Mundurucú society was impressive in the extent to which it maintained cooperation rather than competition (both among and within villages), avoided hierarchy in collective activity, integrated work with sociability, and engaged the entire community in enough diverse collective activities that a sense of commonality was strong.

The Mundurucú provide a model of what is possible in terms of egalitarian cooperative effort, but their specific beliefs, customs, and technologies are obviously far removed from modern material expectations. However, if the Mundurucú and other traditional societies can develop ways of working and living together cooperatively, is it unthinkable that modern societies, using contemporary technological know-how, could do the same? Contemporary employee involvement programs, quality circles, self-managed teams, and "lean production" methods sometimes include steps in that direction, but they also tend to exhibit many democratic limitations.[24] Within modern contexts there are, nevertheless, more fully developed examples of communitarian/cooperative technologies.

Community via Organizational Innovation

This first example illustrates some ways in which the social organization of work can democratically improve without necessarily redesigning tools, machinery, or buildings. During the 1940s and 1950s, approximately 100 "Communities of Work" were established in France, Belgium, Switzerland, and Holland. One called Boimondau is considered representative.[25]

In the early 1940s Marcel Barbu, a worker in a French watchcase factory, decided to see whether an egalitarian and convivial work system could be devised. He and his wife went into business for themselves manufacturing watchcases, beginning with only a few machines. They quickly learned that men who had previously worked in traditional factories were thoroughly socialized into expecting and reproducing hierarchical work relations. Barbu therefore began to hire industrially inexperienced young men off the street, offering to train them if they would assist in his social experiment.

This first innovation was to encourage everyone in the factory to speak out when angry. This system contributed to an initial climate of trust but afterward began wasting time. The workers then decided to hold weekly meetings to air their disagreements. To help establish a sense of community, they also sought to establish a set of shared ethical principles that they would endeavor to respect. Because the group included workers of diverse religious backgrounds and political orientations, the process of reaching consensus was difficult. But it resulted in a set of principles that all pledged they would try to obey.

It next emerged that the factory members, many of whom had had their schooling cut short by World War II, wanted to become more educated. They decided to use any profit gains from their increased productivity to hire teachers and set aside time for courses. Within 3 months, work that used to take them 48 hours took only 36. They began to pay themselves during the time they had saved to participate in group sings and to take courses in grammar, literature, art, business, engineering, physics, politics, religion, dancing, and basketball. In other words, they decided to exchange salary gains and material acquisition for personal and collective self-development.

With time, the group established a merit-oriented salary scale that took into account anything a worker could do that contributed to the group's life, whether as a skilled machinist, a storyteller, a musician, or whatever. Reflecting their ongoing commitment to life beyond the workplace, the community of workers endorsed a principle stating that each member should be "actively related to the whole world."[26] For the pleasure that it would provide, the community also purchased a 235-acre farm, and enabled its workers and their families to spend three paid 10-day periods there during the year.

Two years into the experiment, Barbu and the 90-member workforce decided to reimburse Barbu for his investment in the factory and to turn ownership over to the entire community. Maintaining their weekly meetings, the workers established a system of representative government to manage daily affairs. As the factory grew, the worker-owners organized themselves into work teams of 10 members, several teams into a shop, and several shops into a service. However, the basic 10-member work unit kept daily relations on a friendly, human scale. By the early 1950s the Boimondau watchcase factory was among the seven largest of 50 watchcase factories in France, becoming in later decades an influential model for other communitarian/cooperative experiments.[27]

Boimondau is a modern industrial example of an integrated technological practice analogous in its quality of social relations to that maintained by the Mundurucú.[28] Boimondau did adopt a sort of hierarchy in the form of its elected representative management. But it created a hierarchy grounded in accountability to all the workers, who continued to meet routinely in subgroups and in a semiannual General Assembly. Moreover, members rotated between levels in the hierarchy in response to periodic votes of the entire workforce. Such arrangements seem appropriate for distinguishing democratically legitimate hierarchy from authoritarianism.

The Boimondau experiment relied on the prevailing technical machinery of its day and, within that constraint, created a cooperative style of work that in time transcended the workplace and became a cooperative style of life. No doubt an important factor contributing to

Boimondau's success was that the factory began as a voluntary experiment rather than as an attempt to reorganize an older, traditionally indoctrinated workforce. Moreover, during the initial phase France was under Nazi occupation—an external factor that generated worker solidarity grounded in shared antifascist sentiment. Comparable experiments within established workplaces have frequently experienced resistance from supervisors and managers accustomed to getting their way and from workers, who have learned to be suspicious of the motives behind proposed innovations.[29]

Whether Boimondau might have accomplished more had its buildings, layout, and equipment been altered was not explored. However, in some industries it is clear that the existing physical plant and equipment pose constraints against democratically oriented social reorganization.[30] The following case reveals some of the limitations that may arise when reorganizing work within the context of an established enterprise, but it also illustrates the potentially constructive role that redesigning technical hardware can play in establishing communitarian/cooperative relations.

Community via Technical Innovation

The Volvo corporation has undertaken a series of experiments in workplace reorganization. One of the most ambitious involved constructing a new automobile factory at Kalmar, Sweden. The key hardware innovation was to replace the linear automobile assembly line with a system of independently movable electronic dollies. Each battery-powered dolly transported a single auto chassis, and the dollies could move anywhere within the factory, guided by a centrally controlled computer.[31]

Instead of working on an assembly line, workers at Kalmar were organized into small workshops manned by teams of 15 to 20 members. Each team was responsible for completing the assembly of an entire automobile subsystem on a dolly-mounted chassis. Team members met daily to determine their internal work assignments in consultation with a management specialist. Thus they had some flexibility to vary the nature, order, and allocation of tasks and—within relatively narrow limits established by the management-controlled computer—to vary the work pace.

Volvo management benefitted from reduced employee turnover and absenteeism. Volvo workers enjoyed increased fellowship and the chance to choose and vary their daily tasks. The dolly system permitted more human interaction throughout the day than is possible when workers are stationed along an assembly line moving at a constant pace. Kalmar thus offers an example of one way in which the redesign of machinery and buildings can facilitate more interactive and cooperative workplace relations.

But regarding the overall social restructuring of work, Kalmar seems to have accomplished less than Boimondau and much less than its system of movable dollies could theoretically permit.[32] Kalmar continued to exhibit an authoritarian social control structure. Workers did not participate in designing the plant, had no say in programming the central computer, and played no role in determining broader corporate policy (e.g., the allocation of work among Volvo factories, the range of products to be produced, where and how to market, etc.). The creation of small team-sized workshops presumably facilitated intrateam solidarity, but Volvo management also chose to provide each team with a separate cafeteria, locker room, and plant entrance. Thus the factory's design prevented members of different teams from interacting during the day. Moreover, unlike Boimondau, Kalmar did nothing to establish a sense of community transcending the workplace. Work life and home life remained two thoroughly distinct spheres of activity, such that there was no thrust toward developing a general commitment to advance the broader, transcommunity common good.

Combining Organizational and Technical Innovation

Are there examples that combine Boimondau's commitment to social reorganization with Kalmar-like technical innovations? That would provide the broadest opportunity for establishing equality of power and respect not limited by short-run technical constraints.

A 1980s government program in Zurich, Switzerland, offered neighborhoods legal advice and architectural counseling aimed at increasing community interaction. As a result, neighbors began to remove backyard fences; to build new walkways, gardens, and other neighborhood facilities; and generally to refashion private yards into a well-balanced blend of private, semipublic, and public spaces.[33] Because it was relatively modest in its means and ambitions, this program might be widely adaptable elsewhere in the near term. It also counters the otherwise prevailing sweep toward complementary individualism–authoritarianism.

A housing movement born in Denmark in the mid-1960s seeks, more ambitiously, to integrate desirable aspects of village life with such contemporary realities as urbanization, smaller families, single-parent or working-parent households, and greater sexual equality. The result is "cohousing": resident-planned communities ranging today from 6 to 80 households. More than 100 cohousing communities exist in Denmark and the Netherlands, and the concept is spreading to the United States and elsewhere. (There are now three established cohousing developments in my own small town of Amherst, Massachusetts.)

The Trudeslund cohousing community, near Copenhagen, includes 33 families. Homes for each family cluster along two garden-lined pedestrian streets and are surrounded by ample open space and forested areas. Each home has its own kitchen, living room, and private bedrooms, with rooms somewhat downsized so that the savings can be used to construct and maintain common facilities. The latter include picnic tables, sandboxes, a parking lot and, most importantly, a "common house" with a large kitchen and dining room, play rooms, a darkroom, a workshop, a laundry room, and a community store. Each night residents have the option of eating in the common room; cooking responsibilities rotate among all community adults, so that everyone cooks one evening a month. Because the community is designed to have residents walk past the common house on the way from the parking lot to their homes, the common house becomes a natural gathering spot.[34]

Trudeslund is successful by many measures. The common facilities save time and money, daycare and baby sitting flow naturally from the pattern of community life, social interaction flourishes without sacrificing privacy, and safety and conviviality have both prospered by banishing cars to the community outskirts. Over time cooperation has grown, with resident families choosing to purchase and share collectively such things as tools, a car, a sailboat, and a vacation home. Rather than becoming insular, residents are actively involved in life outside Trudeslund, with the common house serving as a meeting location for transcommunity activities.

Thus, with respect to realizing democratic community, Boimondau (nominally a place of work) and cohousing (an approach to residence) appear able to achieve comparably favorable results.

ALTERNATIVES TO THE PRESENT SYSTEM

Can an alternative communitarian/cooperative style be integrated into an overall system of technologically mediated social relations that would be democratically preferable to today's system?

First, the images drawn here of individualized, authoritarian, and communitarian/cooperative technologies are ideal types, whereas many actual technologies can be intermediate or hybrid. One can imagine, for example, an enterprise that is internally egalitarian, but authoritarian in its relations toward external people and groups. Moreover, between purely authoritarian and purely communitarian/cooperative technologies there runs a complex gradient of intermediate kinds. This suggests that there can be many feasible intermediate steps and alternative end forms in any transition away from authoritarianism.

Second, although authoritarian technologies are unambiguously undemocratic—as is also the present system of complementary authoritarian, individualized, and mass technologies—there is nothing necessarily wrong with permitting a range of individualized technologies. What is important is that these neither represent an essential component of a broader antidemocratic system nor foreclose the availability of a richly diverse array of communitarian/cooperative options. (This explains Design Criterion A's suggestion that it is important to seek an alternative system of relatively communitarian/cooperative and individualized technologies.)

What of mass technologies? Mass technologies are in certain respects authoritarian, but nevertheless democratically ambiguous. They are somewhat undesirable in themselves and thoroughly undesirable given their role within a broader anticommunitarian system. However, they would not necessarily remain unacceptable provided that they were carefully limited, via democratic governance or oversight, in their number and social influence, and thus constituted a relatively marginal component within a sociotechnological order that basically displayed a balance between communitarian/cooperativism and individualism. (The attempt here, within the context of elaborating general democratic design criteria, is to establish guidelines that are as flexible and indeterminate as is sufficient to secure strong democracy.)

In contemplating a transition from an authoritarian technological order, one should not automatically assume that it is necessary merely to replace authoritarian technologies with communitarian/cooperative counterparts, thus leaving intact the present range of individualized and mass technologies. The transition away from authoritarianism might, for instance, also include replacing certain individualized and mass technologies with communitarian/cooperative counterparts.

Moreover, just as communitarian/cooperative technologies can help constitute local democratic communities, so democratically evolved transcommunity technologies would help constitute a democratically federated macrocommunity. Transcommunity technologies could include systems of transportation and communication (consistent with other democratic technological design criteria) and convivial practices such as regional fairs, open-air markets, regional parks, and participatory cultural and sporting events.[35] Community-to-community video teleconferencing, which played a constructive role in 1980s U.S.–Soviet citizen diplomacy, could be a further example.[36]

The U.S. interstate highway system exemplifies a translocal technological network that has frequently proven destructive to local community—and particularly to housewives, poor people, and disadvantaged minority groups.[37] In contrast, many European transit systems have been designed in ways more respectful of face-to-face community life. As with communitarian/cooperativism, the goal in designing transcommunity technologies should extend beyond merely establishing relations of instrumental interdependence toward fostering societywide political equality, mutual respect, understanding, democratic empowerment, or commonality.

A Broader Context

The previous examples of communitarian/cooperative technologies all involved relatively small, self-contained communities. Is the leap from there to wide-scale communitarian/cooperativism merely a matter of massively replicating small-scale models? Not quite. Chapter 3 proposed that societies should comprise a culturally diverse array of communities integrated into a larger, democratically federated political system. Nonauthoritarian technologies are actually easier to envision integrated within such a social order than as isolated democratic oases.

For instance, one barrier to establishing stability among some modern communitarian/cooperative experiments has been that each tries to be a total institution, meeting all of every member's social needs. That puts an extreme performative burden on the institution, creates coercive pressures toward conformity with group norms,[38] and increases the likelihood of chronic interpersonal tension. Moreover, because these experiments represent isolated instances, there is little alternative—if tension becomes truly great—to individuals' migration back to the authoritarian–individualized mainstream or to community dissolution.[39] Meanwhile, the external, relatively authoritarian social environment typically offers little positive support to communitarian/cooperativism and often actively subverts it.[40]

Analogous problems confront communitarian/cooperativism when it occurs not within the context of a larger, authoritarian social setting but on a small-scale, self-reliant basis (as among the Mundurucú). Here the individual may have scant choice but to remain within a single community or another much like it, come what may. This enhances the likelihood of institutional loyalty and stability but may also restrict cultural opportunities or encourage burdensome repression of interpersonal conflict.[41]

Imagine communitarian/cooperativism instead within the context of a broader democratic society.[42] One needn't envision an entire nation, but just a region or even a city. All of the preceding problems might lessen. Living within a democratically self-governed city or town, one that encouraged flexibly schedulable work patterns within a system of semiautonomous communitarian/cooperative workplaces and neighborhoods, each person would have the option of affiliating with a variety of different microcommunities or subcultures.[43] No one institution would be required to meet all of anyone's social needs, alleviating some of the coercion that can otherwise exacerbate conflict and tension. Democratic self-management and job rotation within workplaces would—besides being democratically good in themselves—increase the possibility of flexible work scheduling and diversified careers, because people could more easily fill in for each other. There would be more opportunity for variation and choice in daily activities than in an authoritarian–individualized sociotechnological order, an isolated cooperative workplace, or a small traditional society such as that of the Mundurucú.

Finally, the broader community would, by supposition, support communitarianism and cooperation. For instance, in Spain's Basque region the 100-plus, thriving, worker-owned cooperatives in the Mondragon system are linked together and supported by a complex network of secondary institutions that provide managerial advice, financing, research and development services, health care, housing, and education.[44] As Whyte reports,

> One general conclusion in the Mondragon case is clear: an individual, isolated cooperative in a sea of private enterprises has poor prospects for survival and growth. There is great need to develop a supporting infrastructure.[45]

Depending on community size, and certainly with respect to intercommunity relations, it would generally not be possible for everyone to know everyone else. However, it would be desirable for each person to live or work cooperatively with people representative of many subcultures living in the community. Boimondau's workers adopted this principle when they organized small "Neighbor Groups" of nearby families for periodic self-help meetings.[46]

The relative paucity of examples of long-lived communitarian/cooperative organizations within modern societies certainly suggests that one ought not to underestimate the power of various hindering factors. But those who find in the historical record decisive grounds for dismissing the potential stability of such organizational forms often fail to take into account the existence and historically contingent nature of sundry, unfavorable background conditions; the converse high failure rate among many other organizational forms (such as modern marriages and traditional businesses); and the fact that for tens of millennia longer than any modern institution has yet survived, communitarian/cooperativism appears to have been the norm among our prehistoric, hunter–gatherer ancestors.[47]

Intercommunity Relations

If a self-governed workplace is embedded within a broader democratic community, which in turn becomes part of an even broader democratic federation, who should participate in which decisions? Part I formulated the basic answer: people should have equal and extensive opportunities to influence, directly or indirectly, whichever decisions affect them—especially those that affect their prospects for individual freedom and democracy.[48] This implies that local communities should be empowered to influence those aspects of workplace decisions that would significantly influence the community, such as those involving pollution abatement or plant closings.[49] There also need to be political means of overseeing workplaces' and local communities' compliance with the categorical imperative and of helping to ensure that anything produced remains consistent with democracy (e.g., as specified by all of Part II's proposed design criteria).

The last point is of the utmost democratic significance, yet it is overlooked by virtually all theories of workplace democratization. The custom is to attend to the social quality of the work process—and perhaps to the effects on workers' political consciousness—while ignoring substantive issues concerning the extraworkplace implications of the artifacts and services that are produced. For instance, the concluding section of a deservedly influential essay by historian David F. Noble lauds Norwegian factory workers' involvement in technology choices that helped workers maintain autonomy and creativity. But the essay does not comment on the fact that the plant in question was a government-owned weapons production factory.[50] Why not, instead, ask not only whether a Kalmar (or a General Motors, a Rockwell International, or a Merrill Lynch) is democratic in its immediate social relations, but also whether the automobiles (trucks, fighter planes, or financial services) that it produces are—together with the broader sociotechnological systems they help constitute—compatible with democracy beyond the immediate workplace?

Size Limitations?

Does establishing democratic social relations impose size limits on communities? Probably, although precise limits cannot be specified acontextually. For instance, for a large city to achieve the type of political participation and fellowship that is possible in small communities, urban settlements would probably have to reorganize themselves into federations of smaller, politically semiautonomous communities or neighborhoods. A number of American cities have pursued this strategy for creating more decentralized democracy with relative success.[51] To facilitate the necessary community coalescence, neighborhoods might turn to cooperative economic activities, public amenities, and strategically situated convivial public spaces, and also might employ such devices as external fences, hedges, canals, or other physical means of symbolically demarcating neighborhood boundaries.[52]

What of today's giant, geographically dispersed business enterprises and public bureaucracies? At a minimum, they too should be reorganized into federations of quasi-autonomous, democratic divisions and subsidiaries.[53] In fact, many already are organized into appropriate subdivisions and some have even begun to significantly decentralize managerial control and strategic planning,[54] although as yet the qualification that each subsidiary and the whole should be democratically governed has generally not taken root.

Reorganized this way, giant corporations and government agencies might be democratically permissible but probably still would be far from optimal. In other words, to the extent that democratic community can be realized most fully on a small or intermediate scale with ample face-to-face interaction—so that people can know one another in context and thus most completely—a supplementary guideline would be that organizations ought ordinarily to be no larger in membership, and no more geographically dispersed, than is required to produce a democratically necessary or desired good.

If one believes that modern organizational gigantism reflects crucial economies of scale, this second requirement will carry little force. But there is considerable evidence that there are many organizational and technical means to produce both humanely and economically at scales vastly smaller than today's norms.[55] In north-central Italy, for instance, cooperative networks of small manufacturing firms design and produce cutting-edge products that sell competitively in world markets. Of Italy's 20 regions, Emilia-Romagna is the fastest growing and has the highest per capita income. In 1980 nearly 60 percent of the region's 1.7

million workers were employed in manufacturing, but 9 out of 10 manufacturing firms employed fewer than 100 workers.[56]

To the extent that immense size can reflect little beyond the quest for democratically unaccountable power, doesn't it seem reasonable to ask bigness advocates to bear the burden of proof?

VIRTUAL COMMUNITY?

Contemporary technological reporting is rife with notions of electronically mediated communities transcending any sense of spatial confinement.[57] Could "virtual communities"—in which, for example, people can interact electronically with a "community" of fellow citizens spread across entire regions or continents—replace territorially situated social relations? It would be premature to give a definitive answer.

Note, first, that telecommunications technologies encompass a broad and evolving range of systems. Second, the social dimensions of cable television, electronic mail, fax machines, and so forth remain in many respects indeterminate until the particulars of context, system design, and the community of owners, operators, users, and other affected parties are all specified.[58] Third, the world is saturated with many kinds of technologies, and one should therefore resist presuming that the social significance of any one category (such as telecommunications) will automatically overshadow the rest. After all, at Boimondau and Trudeslund no single practice or design innovation was decisive in establishing democratic community. Fourth, it is important to evaluate telecommunications technologies, like any technology, using a comprehensive set of democratic design criteria, not just one (see the remainder of Part II).

Notwithstanding these remarks, there seem to be context-invariant reasons for suspecting that virtual communities should probably not replace territorially localized communities as the foundation of democratic polities. One can reach this conclusion (1) by attending to technologies' focal and nonfocal dimensions and (2) by evaluating these dimensions using Chapter 3's criteria for democratic community.

First, electronically mediated communication filters out and alters nuance, warmth, contextuality, and so on that seem important to fully human, morally engaged interaction.[59] That is one reason why many Japanese and European executives persist in considering face-to-face encounters essential to their business dealings and why many engineers, too, prefer face-to-face interaction and find it essential to their creativity.[60] Second, there is much concern that screen-based technologies (such as television and computer monitors) are prone to induce democratically unpromising psychopathologies, ranging from escapism to passivity, obsession, confusing watching with doing, withdrawal from other forms of social engagement, and psychological distancing from moral consequences.[61]

Third, to the extent that membership in virtual communities proves less stable than that which obtains in other forms of democratic community, or that social relations prove less thick (i.e., less embedded in a context saturated in shared meaning and history), there could be adverse consequences for individual psychological and moral development. In the words of psychologist Robert Kegan:

> Long-term relationships and life in a community of considerable duration may be essential if we are not to lose ourselves, if we are to be able to recollect ourselves. They may be essential to the human coherence of our lives, a coherence which is not found from looking into the faces of those who relieve us because they know nothing of us when we were less than ourselves, but from looking into the faces of those who relieve us because they reflect our history in their faces, faces which we can look into finally without anger or shame, and which look back at us with love.[62]

Fourth, no matter with whom one communicates nor how far one's imagination flies, one's body—and hence many material interdependencies with other people—always remains locally situated. Thus it seems morally hazardous to envision communing with far-flung telemates, if that means growing indifferent to physical neighbors.[63] It is not encouraging to observe just such indifference in California's Silicon Valley, one of the world's most "highly wired" regions.[64]

Fifth, one function of democratic community is to provide a social foundation for egalitarian empowerment within a process of democratic structuration. This suggests that local community boundaries ought normally to remain roughly contiguous with the territorial boundaries defining formal political accountability and agency. Yet the criterion of local self-governance will be breached if involvement in spatially dispersed social networks grows to subvert a collective capacity to govern the locales people physically inhabit.[65] And the criterion of egalitarianism will be breached if coveys of technorich cronies are empowered to telelobby senators, while their technopoor neighbors are excluded from the circuit.

Telecommunications enthusiasts, such as contemporary boosters of a national information "superhighway," sometimes respond that should a mismatch arise between bonds of social affiliation, which could

increasingly become nonterritorial, versus current political jurisdictions, "political systems can change."[66] However, that answer provides
none of the necessary specifics. It also fails, for example, to grapple with
the U.S. Constitution's requirement that amendments garner the support of a majority of elected federal or state legislators. How readily do
legislators normally accede to voting away their own offices? In short,
at a minimum, one would confront a profound transition dilemma.

If the prospect of telecommunity replacing spatially localized community ought to evoke skepticism or opposition, one can nevertheless
remain open to the possibility of democratically managing the evolution
of telecommunications systems in ways that instead supplement more
traditional forms of democratic community. Caution is in order. However, the benefits can potentially include combating local parochialism;
helping to establish individual memberships in a diverse range of
communities, associations, and social movements; empowering isolated
or marginalized groups; and facilitating transcommunity and intersocietal understanding, coordination, and accountability.[67]

Systems designed to support such uses—especially without subvert-
ing local community—are unlikely to emerge without concerted democratic struggle.[68] For instance, one transnational corporation first supported, then clandestinely monitored, and finally terminated a
successful, productivity-enhancing internal computer conferencing system. Senior managers had discovered that their subordinates, including
women executives previously isolated from one another, were spending
coffee-break time on the system discussing and criticizing company
policy.[69] Moreover, even seemingly benign systems require ongoing
scrutiny for psychodevelopmental and social structural effects that may
only emerge gradually as a system evolves.

CONCLUSION

A transition from authoritarianism is technically and organizationally
practicable, but current circumstances militate against this transition
being effortless. No one can simply engineer or legislate mutuality into
existence. Much prevailing technology and architecture function materially and ideologically against democratization. These and other
factors work against democracy (technological and otherwise) but do
not preclude it. Many technologies, although presently designed or
deployed in an authoritarian fashion, harbor latent democratic potentialities.

Movements toward communitarian/cooperativism are already

afoot in many quarters. Each of the more successful examples that we have considered no doubt benefited from special local conditions (e.g., shared antifascist sentiment in the case of Boimondau and a strong labor movement and social democratic policies in Scandinavia). However, the nondetermining nature of social structures, and their partial embodiment in human belief systems and practices, suggests that local opportunities for democratization may actually be more widespread than the phrase "special conditions" might otherwise imply.[70]

Chapter 5

"NO MAN IS SACRIFICED TO THE WANTS OF ANOTHER"
Democratic Work

Beyond democratic community, another necessary condition of democracy is that citizens have extensive opportunities for self-actualizing experience, in the form of democratic work (see Chapter 3). However, Chapter 2 noted that technologies, as one aspect of their polypotency, shape and help constitute possibilities for human action and psychological development. Insofar as (1) citizens should have equal and extensive opportunities to develop and express moral autonomy and (2) technologies structure patterns of opportunities and constraints for human endeavor and psychological development, it is (3) democratically vital that technologies' design and social organization support equal and extensive opportunities for people to develop and express moral autonomy.

Chapter 3 began elaborating one aspect of this recommendation by prescribing opportunities for flexible life scheduling. With respect to technology, it is thus democratically important to:

> Seek equal and extensive availability of a diverse array of flexibly schedulable, self-actualizing technological practices. Minimize the social need for meaningless, debilitating, or otherwise autonomy-impairing technological practices.

This is Design Criterion B.

Often, social criticism of technology conflates a concern for self-actualization with that for democratic community, implying that technologies are either (1) authoritarian and (for subordinates) dull, dangerous, or degrading, or else (2) egalitarian, convivial, and self-actualizing. But this can be misleading. For instance, communitarian/cooperative technologies do incline toward compatibility with self-actualization, because they lack autonomy-crippling hierarchy and afford creative opportunities for self-governance. However, one can also conceive of egalitarian convivial practices in which, apart from the pleasures of sociability, the work is nonstimulating. Imagine, for example, spending days sitting around a table and chatting with close friends while tearing pieces of paper into confetti strips.[1]

Self-actualization is morally good in itself. But it can also contribute to democratic institutional stability (insofar as institutions earn citizens' respect by nurturing their autonomy) and to commonalities (insofar as people come to regard the self-actualizing activities of others as enriching the overall social fabric).[2] As one example, recall the Boimondau workers, who paid one another to develop themselves in any way that contributed to group life.

CONTEMPORARY TECHNOLOGICAL ORDERS

The ideal of self-actualization contrasts markedly with the reality that has prevailed throughout most of modern history. The examples discussed in this chapter focus on industrial occupations, but a similar case can be made with regard to other sorts of work, including household labor and schooling, as well as ancillary activities such as commuting, shopping, and mass recreation.[3]

Throughout industrial societies, a hierarchically graded division of labor and an extreme subdivision of tasks has frequently stripped lower-echelon workers of significant opportunities to work creatively. Adam Smith identified this tendency two centuries ago:

> In the progress of the division of labour, the employment of . . . the great body of the people, comes to be confined to a few very simple operations, frequently to one or two. . . . The man whose whole life is spent in performing a few simple operations . . . has no occasion to exert his understanding, or to exercise his invention in finding out expedients for removing difficulties which never occur. He naturally . . . becomes as stupid and as ignorant as it is possible for a human creature to become.[4]

Taken to its worst extreme, technological practice not only blocks opportunities for self-development, but positively negates workers' sense of themselves, even their choice of how and when to move their bodies. This is most obvious in the case of assembly-line production, but Marx observed it earlier in conjunction with large-scale water- and steam-powered machinery.[5] By the early 20th century the drive to enhance productivity by making work dull and unchallenging became enshrined as a virtue by Frederick W. Taylor, the father of "scientific management":

> Under scientific management . . . the managers assume . . . the burden of gathering together all of the traditional knowledge which in the past has been possessed by the workmen and then of . . . reducing this knowledge to rules, laws and formulae.
> Without the element of stupidity the method would never work. A pig iron handler shall be so stupid and phlegmatic that he more resembles the ox than any other type. . . . The man who is mentally alert is entirely unsuited to what would, for him, be the grinding monotony of work of this character.[6]

Similar theories were gradually extended to the organization of offices and other nonfactory work environments and most recently— especially with the proliferation of computers and microelectronics— have begun to prey on the autonomy of white-collar workers such as midlevel managers, architects, computer programmers, engineers, and industrial designers.[7] A study of several hundred workers in one U.S. job market concluded that "most of them expend more mental effort and resourcefulness in getting to work than in doing their jobs."[8]

Even by orthodox economic standards, the results have often proven counterproductive. Workers who might otherwise channel their energy into creating useful and pleasing products, instead exert themselves developing ingenious methods of eluding oppressive work routines.[9] By higher order moral and democratic standards, the results have been tragic: "citizenries" laden with people whose skills are stunted, whose morality is uncritically conformist rather than self-guided, who shun creativity and challenge, who are fatalistic about the possibility of desirable social change, who value submission to authority more than personal autonomy, and who transmit these democratically adverse traits and expectations to their children.[10]

During the past 15 years, it has become increasingly clear that there are technically and economically viable alternatives to the most oppressive extremes of Taylorist mass production. For instance, analysts of industrial relations have identified a range of approaches for more flexibly producing smaller and more diverse product ranges, in part by devolving selected managerial and maintenance responsibilities to teams of shop-floor workers.[11] However, it is also clear that these alternatives can readily come in authoritarian or uncreative variants.[12]

For example, Japanese auto manufacturers have developed a high-quality/low-inventory mass production system known as "lean production" or "just-in-time" (JIT). Some JIT plants achieve high productivity in part by deploying supervisory techniques that greatly intensify the pace at which workers are compelled to perform.[13] However, even the most humane realizations of JIT have drawbacks: the well-regarded NUMMI (New United Motor Manufacturing, Inc.) auto plant in California, a joint venture of Toyota and General Motors, has teams of assembly workers participating in preventive maintenance and quality assurance, while rotating jobs internally. But basic assembly tasks remain unchallenging (e.g., a job cycle is finished in about 60 seconds). The workers are also trained to conduct scientific time-and-motion studies of their own work processes, resulting in minutely specified, management-enforced rules governing their moment-by-moment bodily motions. Sympathetic observers characterize this as an example of "democratic Taylorism."[14]

The historical origin of widespread work degradation is complex and, within this book's context, significant only in the sense that it is crucial to understand it as the result of contingent historical choices and forces rather than as technologically inevitable.[15] This is encouraging insofar as it indicates that improvement is possible, even if the extent of that improvement is apt to remain limited without principled democratic effort.

Adopting a somewhat specialized division of labor can enhance labor productivity—in the orthodox sense of increasing the quantity and diversity of material goods and services produced by each worker—if that is what the members of a society desire. (This definition of productivity is "orthodox" because it measures productivity in terms of focal or monetary results but neglects nonfocal results, such as the subjective quality of the work experience, its psychodevelopmental effects upon workers, and the consistency of what is produced with democracy and with communities' other common concerns.[16]) If a task requires carefully timed or complex interaction among a number of different people, it can also be helpful to create the role of work coordinator.

This kind of division of labor does not compel each person to be occupied with a single limited sort of work. Most jobs can be challenging and can afford a range of creativity. It is also possible to share, rotate, and shift jobs (within and among workplaces) to further enhance chances for self-development and autonomy.

In contrast, ultraspecialization and more extensive and rigid work hierarchies have often been adopted for reasons related at best indirectly to productivity enhancement. New machinery has in some cases been used to replace skilled workers with unskilled ones, to prevent communication among workers, to monitor and control the work pace, and to standardize output—perhaps lowering prices, but at the expense of diminished opportunities for creativity and of diminished availability of a diversified range of creatively crafted products and services.[17] Sometimes productivity improved, but the number or quality of jobs declined. Other times productivity was little altered, but wages were bargained down. And in many cases neither productivity nor profits improved. Thus by all standards, except perhaps managers' power to manipulate workers, work has been degraded for no remotely rational reason.[18]

From the viewpoint of strong democracy, it is vital to challenge any social structure that sacrifices opportunities for self-actualization to corporate profits or to reproducing illegitimate hierarchy. (There need be nothing democratically wrong with profits per se; the problem is when democracy and moral autonomy become profits' sacrificial victims.) When it is possible to improve opportunities for self-development and creativity without sacrificing productivity or total output, that improvement should without question be undertaken. Even when a trade-off is unavoidable, the appropriate standard should be to favor self-actualization and democracy generally. The exceptions would be when it is persuasively counterargued and democratically affirmed that (1) the consequent decline in economic output translates into an offsetting structural diminution in democracy or freedom or (2) citizens' informed preference is for higher output over an enhanced quality of work.

The latter qualifications concede less than might be supposed. First, "democratically affirmed" means that the background conditions necessary for informed, egalitarian, and participatory deliberative procedures need to be in place. That entails, among other things, the prior existence of work opportunities allowing citizens to know what it is like to live and work convivially, creatively, and with a level of responsibility and challenge commensurate with their abilities and aspirations. Second, the social burdens—in terms of diminished opportunities for self-actualization—resulting from any democratically sanctioned trade-off ought to be equitably shared.[19] This qualification is extremely important, but historically it has rarely been observed.

Third, to the extent that communitarian/cooperative and self-actualizing technologies have already replaced authoritarian and self-negating ones, the chance diminishes of there being a trade-off between quality-of-work versus productivity. This is because the present radical dichotomy of work versus leisure or consumption would partially dissolve, in effect enlarging the ends toward which work and productivity would be aimed. Recall, for example, Chapter 4's account of "work" among the Mundurucú or, again, the Boimondau workers' decision to pay themselves to take courses or to practice playing musical instruments.

It is commonplace to fall short of the preceding standards. For instance, because they rate its productivity twice as high, business management professors Paul Adler and Robert Cole favor the just-in-time NUMMI auto plant over Volvo's advanced Uddevalla factory (which they acknowledge offers much more creative work).[20] They reason that if the productivity gap were smaller,

> human advantages would tip the scale in Uddevalla's favor. . . . It would be wonderful if we lived in a world where every job could be an opportunity for Maslovian self-actualization. But when products are fairly standard and mass produced . . . then efficiency requires . . . a form of work organization that precludes . . . very high intrinsic work satisfaction. . . . Any community that needs such standardized goods will object to paying . . . exorbitant costs.[21]

In favoring NUMMI, Adler and Cole are willing to sacrifice one group of workers' opportunities for self-development in return for material benefits that accrue primarily to other people. This violates the democratic moral imperative not to treat some people as instrumental means to other peoples' ends. The workers have not been offered a chance to make an informed choice themselves between alternative forms of labor. No effort is made to ascertain whether NUMMI's high productivity contributes positively to democratizing society's basic structure. There is no democratic process in place for allowing communities to directly register their preference for freedom as realized in creative work versus freedom in consuming a greater abundance of inexpensive, mass-produced goods. The supposedly urgent need to increase productivity is occurring during a time of deepening structural unemployment and underemployment, which means that human resource utilization societywide is far from full capacity.[22] The point is not that Adler and Cole's judgment in favor of NUMMI is necessarily wrong, but that their reasoning is democratically deficient.

SELF-ACTUALIZING ALTERNATIVES

As in shifting toward a more communitarian/cooperative technological order, creating self-actualizing technological practices can entail innovations in social organization or in the design and arrangement of artifacts and buildings. That this is feasible—presuming that the will to innovate exists—is becoming widely appreciated.[23] Obvious social innovations can include job enlargement or, as at NUMMI, job rotation. If, however, a worker who formerly performed a single monotonous task now performs a number of such tasks in succession, the degree of improvement is slight. Instead of being bored at one job, each worker spends the day moving from one unchallenging and uncreative job to the next. More meaningful change only occurs if individual tasks are redesigned to enhance their intrinsic interest or creativity and especially if workers succeed in recovering power to help decide what and how to produce.

Prior to World War II, for instance, operating a machine tool was a high-skill job. Since then, the gradual introduction of numerically controlled (i.e., programmable) machine tools has permitted managers in some shops to downgrade the job of machine tool operator to positioning a piece of metal that needs to be shaped, pushing a button, and watching to make sure nothing goes wrong. The machine does the shaping, following instructions previously entered by professional programmers. In short, a job that once required considerable dexterity, judgment, and ingenuity on the part of a machine tool operator can now be accomplished in ways that afford shop floor personnel diminished opportunities to develop and exercise their skills.

However, some machine shops reveal alternative possibilities. For instance, in Norway, a powerful union, together with social democratic legislation that supports worker participation in designing work conditions, began to make programming numerically controlled machine tools part of the operator's job nearly two decades ago. In one state-owned factory, workers used a built-in microprocessor to prepare and revise the paper tapes guiding their machines. Such work is different— and conceivably worse[24]—than it was prior to the introduction of numerically controlled machines. But it is unquestionably more challenging than the work undertaken with the same machines in many metalworking shops in the United States and elsewhere. Thus, this is a case in which similar equipment can, according to the social context of its use, support either worker self-development or tedium and subordination.[25] Furthermore, the opportunity for meaningful, creative effort can much increase if a firm produces a diverse or changing range of products, approximating more closely the rewards of craft production.[26]

Many examples of changes in the design of productive equipment that permit work to be more challenging and creative can be cited. The Volvo auto factory at Kalmar permitted not only cooperative group effort but also flexible group-determined production methods.[27] As opposed to assembly-line work in which a highly specialized division of labor is embodied in the machinery itself, the Kalmar layout allowed work teams to modify their internal division of labor by rotating jobs and restructuring their division of tasks within relatively broad constraints. These constraints were, moreover, determined more by management-imposed prerogatives than by the limits of what the plant's equipment permitted. Shell International provides another example. When designing a new factory control system, Shell engineers were instructed not to automate all processing stages but instead to ensure that operators would continue to exercise judgment during their work.[28]

The Volvo and Shell technical examples each exhibit limitations in the extent to which corporate management was willing to permit workers to choose the content and conditions of their work. These cases nonetheless support the claim that where there is a social will, there is also ample technical opportunity to redesign equipment and buildings in the interest of permitting work to become more creative.

Automation

One significant issue bearing on both self-actualization and democratic community concerns the social role of automation. Scholars have criticized inhumane versions of automation, formulating at least three alternative ideals: (1) automation as boring or stressful labor that nevertheless, by shortening the average work day, increases the opportunity for satisfying leisure; (2) automation as a form of self-actualizing cooperative labor; and (3) restricted use of automation to eliminate only meaningless or backbreaking labor so that all remaining work would be more humanly satisfying. Many proponents have appreciated that realizing any one of these ideals is apt to require concerted democratic struggle.

The mature Karl Marx propounded the first ideal. Under capitalism, he argued, mechanization and automation displace some workers while forcing those who remain to work long days at a breakneck pace. Within a socialist or communist society the same machinery would, Marx prescribed, be used to shorten the work day so that each worker would have ample free time for self-development through diverse leisure activities.[29] However, if leisure time in a postcapitalist society would be abundant and self-actualizing, work within factories would remain boring at best. According to Marx, necessary work would be made as humane as possible, but it none the less still remains a realm of necessity. Beyond it begins that development of human energy which is an end in itself, the true realm of freedom, which, however, can blossom forth only with this realm of necessity as its basis.[30]

Industrial sociologists began articulating a second ideal during the 1950s. Although it had been agreed that a shift from craft production to factories and mechanization tended to result in less satisfying and unconvivial work, researchers discovered that continuous-process automated factories eliminate routine jobs but create a need for teams of skilled workers to diagnose, repair, and maintain complex machinery. This led to the hypothesis that in a highly automated society many jobs could be of this creative, repair-team variety.[31]

The third ideal foresees a more limited use of automation within the broader context of redesigning products, jobs, and working conditions to make all work both intrinsically and instrumentally satisfying. This is well articulated in the 19th-century writings of William Morris. As one of his characters in a futuristic novel explains:

We have now found out what we want, so we make no more than we want; and as we are not driven to make a vast quantity of useless things, we have time and resources enough to consider our pleasure in making them. All work which would be irksome to do by hand is done by immensely improved machinery; and in all work which it is a pleasure to do by hand machinery is done without. There is no difficulty in finding work which suits the special turn of mind of everybody; so that no man is sacrificed to the wants of another. From time to time, when we have found out that some piece of work was too disagreeable or troublesome, we have given it up and done altogether without the thing produced by it.[32]

For our purposes, important questions concerning automation include: Which of the preceding ideals are technically realizable and under what social conditions? Are they equally desirable? Appropriately implemented, all three ideals appear to be democratically acceptable. And although they are not mutually exclusive, in general the second is preferable to the first, while the third may be best of all.

A problem with the first is that it is not fully consistent with Marx's own earlier criteria of unalienated labor—criteria that Chapter 3 implicitly adopted as partially representing an ideal standard of democratic work. The young Marx envisioned that humane work would be satisfying because one helped choose the means and ends of one's labor, thus developing oneself while creating products that satisfy genuine needs and worthy desires.[33] But Marx's later, mature conception of socialist or communist work fails to meet this standard. Work satisfying real human needs would be relatively uncreative, specialized factory labor, whereas truly creative human endeavor would occur only during leisure hours, and therefore would often comprise unnecessary play or hobbies. Although this plan could conceivably amount to a work day that effectively balances necessary labor against creative play, it might just as conceivably result in alternating between the boredom of routine specialized labor and the self-trivialization or disengagement of mere hobbyism.[34]

The second ideal is an improvement, because it envisions a world in which necessary labor can be creative. But this ideal, especially as originally articulated, harbors problems of its own. First, it may not be realizable. There is some evidence that when first constructed, automated plants do provide opportunities for creative maintenance. But after an initial shakedown period, maintenance can become boringly routine.[35] Second, when work processes are automated in expensive, highly integrated ways, maintenance can turn brutally stressful rather than playfully challenging.[36] Third, even if one envisions ways of circumventing these drawbacks and anticipates that workers will eventually win the right to help determine broader corporate strategy, creative repair and maintenance—if it constitutes the bulk of a worker's productive activity—is unlikely to prove as rewarding as designing and creating new goods or services.

On the other hand, more recent variants of the second ideal foresee promising potentialities in certain computer-based technologies. Methods such as "flexible specialization" or "human-centered computer-integrated manufacture" are technically compatible with deploying teams of highly skilled workers to both program and oversee complex production machinery in response to swiftly evolving market shifts.[37] However, notwithstanding enthusiasts' glowing imagery,[38] flexible computer-customized production seems rather far removed from the sensuous process of individually crafting a custom-made product using hand tools. It is, moreover, possible that the very productivity of such systems, coupled with their applicability to manufacture for specialized market niches, will restrict the percentage of desirable jobs they can provide.[39] Finally, to the extent that flexible specialization is interdependent with mass production (e.g., for some of its inputs),[40] concern could arise that it is morally parasitic upon the degraded labor of others.

Of potentially greater applicability is the vision—dubbed "informating"—of taking the same computers that now automatically guide machinery or manipulate data and using them simultaneously to distribute comprehensive information concerning the work flow to workers throughout a plant, office, or firm.[41] Ideally implemented, computers that otherwise degrade skills, promote isolation, or help monitor and pace workers' performance would instead empower a suitably trained workforce to participate in comprehensively informed, managerial-style decision making.

There may be several residual democratic concerns with this otherwise quite appealing vision. One is that computer-mediated sociability may prove less satisfactory than face-to-face interaction.[42] Second, informating envisions a definitive shift from tactile, craft, and other embodied skills to more purely intellectual labor. Some people may, however, be reluctant to relinquish the satisfactions they find in bodily engagement.[43] In that case, democratic governance procedures ought presumably to ensure that sufficient job opportunities of this sort remain available. Moreover, it is also possible that certain bodily skills and engagements, beyond affording pleasure to some, are helpful or even necessary to moral growth. Modern Western moral psychology appears not to have contemplated this possibility, but it is striking that many Eastern traditions such as Hinduism, Buddhism, Taoism, and Sufism are united in seeing physical practice as essential to spiritual–moral development.

The third ideal—carefully targeting automation to replace jobs that are socially necessary but unavoidably boring or dangerous, while otherwise maximizing opportunities for self-development—escapes the previous problems. If this would result in an economy of diminished aggregate output, it should (as noted earlier) be a matter of broad democratic choice how much opportunity for self-actualizing work to sacrifice in the interest of increased material output. However, too often this ideal is simply ignored.

Doris Lessing, on the other hand, implies in her novel *The Sirian Experiments* that the moment any work can be automated, it becomes humanly unsatisfying and so should be automated.[44] But automation can never exactly duplicate the process or results of skilled craft labor. The craft worker displays skills, nuances, and creativity that machines cannot, and he or she can also make fine adjustments in the work process according to minute variations in tools, materials, and the eventual product's envisioned context of use. Even human "imperfections" can be a source of deeper meaning in the product.[45] This suggests that automated and semiautomated work will not generally result in products or services as humanly satisfying, nor in workers as morally autonomous and fulfilled, as can democratically situated, skilled craft labor or the practice of an art or a profession.

Without being definitive, this discussion should indicate the merit in ensuring that the future of automation is decided within the context of striving to establish structural conditions for strong democracy.

A BROADER CONTEXT

Is it plausible to imagine that anyone who chose to could work part-time, flexible hours within a variety of different kinds of work settings, and yet necessary things would still get done? Don't many technological activities require close collaboration among teams of specially trained individuals? If a woman—for example, Denise—spends Wednesdays working as a truck mechanic, is apprenticing part time as a paramedic, and takes weeks off in the summer to work on a farm, how is the office where she is supposed to be a senior coordinator going to get along without her? Part of the answer can involve designing artifacts and practices adapted to use within an overall social context of flexible career patterns. However, much of the answer is already implicit in Design Criterion A; this begins to reveal the potential for fruitful complementarities within a system of democratic design criteria.

Recall some of the preceding chapter's prescriptions: workplaces, associations, communities, and other common endeavors should be democratically self-governed. To reduce hierarchy, one might provide opportunities for people to upgrade or diversify their skills, thus permitting job rotation among workers and managers or, preferably, more egalitarian and effective collective self-management. Chapter 4 justifies such recommendations on the grounds that they would contribute to establishing democratic community.

That is not the only possible justification. For example, upgrading skills, rotating jobs, and practicing collective self-governance each provide opportunities for self-actualizing experience. Moreover, when communities and workplaces are democratically self-governed (as at Boimondau), or the people within them diversify their skills and rotate jobs (as on a kibbutz or at Kalmar and NUMMI), people (1) tend to evolve a more comprehensive understanding of each overall endeavor in which they are involved and (2) also develop more overlapping ranges of skills and enterprise-specific knowledges.[46]

Result (1) enlarges the meaning that people can derive from their contributions within a collaborative process, which is important to self-actualization.[47] Results (1) and (2) together mean that no one individual, as the unique performer of a specialized role, is indispensable to any given enterprise. Thus rather than being known as the relatively monodimensional performer of a limited set of specialized social roles,[48] individuals could instead become respected as the unique, infinitely faceted, whole evolving beings that they really are. Coworkers could learn to appreciate Denise for her unique self and the inimitable personal qualities she brings to their common endeavors. But when she chooses to be elsewhere, they might learn to appreciate the different skills and style of those who carry on, in their own unique ways, in her absence.[49]

Thus conditions of communitarian/cooperative enterprise (Design Criterion A) facilitate establishing systems of self-actualizing technological practices (Design Criterion B). Conversely, the self-actualization that results from satisfying Criterion B promises to contribute to individuals' capacities to participate effectively—from the standpoint of skills, knowledge, creativity, self-respect, and greater readiness to advance the common good—in communitarian/cooperative endeavors.

Two more questions arise about flexible, diverse career paths. First, is Denise going to experience rewarding self-development or stressed-out dissipation and fragmentation? That depends partly on Denise, partly on other contextual considerations. People need to be able to experiment to discover for themselves a satisfying personal balance. For example, some may prefer to pursue a single, even a simple kind of work, perhaps because they experience it as a form of prayer or meditation, or because they find their challenges and satisfactions in life elsewhere.[50] Alternatively, many modern Israeli kibbutzim appear to have prevented fragmentation by finding career diversity and job rotation within the context of a relatively small, integrated community setting.[51] There a Denise's sense of psychological integration is facilitated by many community members knowing her in all the different roles she performs. Farmers and housewives in the Old Order Amish economy likewise avoid fragmentation because, as well-rounded generalists, their diverse competencies find integrated meaning by contributing visibly to household, farm, and community economic health—all subordinate, in turn, to an overarching religious purpose. Such examples suggest that democratic community (Criterion A) can help offset the risks of psychological fragmentation that might otherwise result from single-minded preoccupation with Criterion B. Of course, Denise cannot ensure on her own that she will have the freedom to choose among a broad array of alternative work patterns. Securing that choice on a general basis will typically require a collective democratic capacity to influence the overall structure of available work opportunities.[52]

Second, do job rotation and job sharing require that individual tasks be redesigned to make them easy to learn?[53] That is a viable, democratically acceptable option. However, it is second best to vastly broadening opportunities for lifelong learning, part-time apprenticeships, and other means of allowing people to enhance their capability to undertake new, complex tasks. The latter strategy is preferable because it contributes to reducing hierarchy without compromising opportunities for self-actualization.

Finally, viewed as an emergent system, Criteria A and B in a sense transcend both of their intended aims (i.e., democratic community and democratic work), promising to contribute synthetically to citizens' general political competency, creativity, and knowledgeability. Competency and creativity would result simply from the application on a broader field of capacities developed within local, self-actualizing, or communitarian/cooperative endeavor.[54] Political knowledgeability would originate from participating in a broader range of societal activities and from interacting with a broader range of people, collaborative endeavors, and communities. Moreover, because technologies are themselves social structures, the opportunity to participate in a wider range of technological endeavors represents one essential means of implementing the basic democratic requirement that citizens have opportunities to participate in governing their societies' social structures.

TECHNOLOGY AND MORAL DEVELOPMENT

Advocates of meaningful or democratic work normally take it for granted that if work is democratically self-managed, varied, and creative, then one has satisfied all relevant considerations bearing on the humanization of technology generally and the work process specifically. But from the viewpoint of strong democracy and freedom, this is not so.

In the first place, technology's democratic significance far transcends its bearing on the work process.[55] Second, it is up to citizens themselves to decide the "relevant considerations." Third, this chapter is not concerned exclusively with work humanization. It is concerned centrally with developing and expressing moral autonomy. There is substantial overlap between those two concepts, but they are not identical. Work humanization connotes such things as dignity, conviviality, intrinsic satisfaction, and creativity. These can each advance or express moral autonomy. However, the latter must also issue specifically in progressively greater individual moral competency, including a readiness to act effectively on behalf of common concerns.

However, much remains unknown about the process of moral development, and there is variability in the process—and hence in its supporting social conditions—among cultures, individuals, and even individuals at different stages in their lives.[56] This suggests, in turn, that much remains to be learned about how technologies influence moral development. Certainly, opportunities for creative self-expression and collective self-governance are generally crucial to developing and expressing autonomy. But creativity remains a somewhat indeterminate concept,[57] and whereas creativity and self-governance are helpful to moral autonomy, they are probably not sufficient.

Consider a few possibilities: First, Western developmental psychologists have tended to focus on cognitive development. Consequently, much less is known (in the West) about emotional development and even less about the specific role of emotional development within moral development.[58] Technologies, on the other hand, obviously exert profound emotional impact—not only when people work with them as practitioners, but also because they constitute a substantial portion of our everyday environment.[59] Perhaps that emotional impact plays an important role in individual moral development. If so, this fact ought to play a fundamental part in technological design and decision making.[60] Certainly, one might ponder such things as the developmental implications of the extremely obsessive and compulsive behavior that interaction with computers elicits from some people.[61]

Second, various design trends may bear some responsibility for heightened psychological fragmentation. For example, many modern technologies—ranging from eyeglasses to telephones, television, and scientific instruments—enhance one dimension of people's immediate sensory experience at the expense of others.[62] Might this tend to elicit fragmented development of perceptual capacities, perhaps with adverse implications for holistic perception, intuition, and synthetic reasoning?[63] If the answer is yes, and if this in turn bears on individuals' moral development, then that consideration too ought to play a fundamental part in decisions concerning technology.[64] Likewise, in the ostensible interest of efficiency, convenience, or safety, many contemporary technologies experientially separate ends and means that were formally integrated within holistic local practices or forms of life.[65] Could that experiential separation contribute to the observed modern eclipse of moral reasoning by instrumental rationality?[66] If so, that would have crucial implications for democratic technological design. The same would also be true if technologies such as personal computers contribute to a morally problematic deepening of the experiential split between emotion and reason.[67]

Third, what if the human meaningfulness of our overall environment—or the dearth of meaning many people experience in a sociotechnological order substantially generated through industrial or bureaucratic production—should turn out to play a significant role in human moral development?[68] A subsidiary issue of meaning involves the plasticity of modern and "postmodern" technological materials. Older forms of craftsmanship, such as woodworking, weaving, pottery making, or sculpting, involve working collaboratively and sensuously with natural materials. There is a kind of reciprocity between maker and made; a respect for the integrity or recalcitrance of wood, rock, or natural fiber; and a related receptivity to what the material can and cannot do, or disclose, or say to us.[69] In contrast, modern malleable materials such as glass, plywood, steel, concrete, plastic, alloy, advanced ceramics and composites, and engineered molecules and genes seem to reflect a progressive increase in the ability to fully impose human willpower on nature. Could this, in turn, jeopardize human moral abilities to hear what the natural world has to say back?[70]

All of the preceding considerations are among the kinds of questions that a democratic politics of technology must learn to address.

Toward DEMOCRATIC COMMUNITY: (Chapter 4)

A.   Seek a balance among communitarian/cooperative, individualized, and transcommunity technologies. Avoid technologies that establish authoritarian social relations.

Toward DEMOCRATIC WORK: (Chapter 5)

B.   Seek a diverse array of flexibly schedulable, self-actualizing technological practices. Avoid meaningless, debilitating, or otherwise autonomy-impairing technological practices.

Toward DEMOCRATIC POLITICS: (Chapter 6)

C.   Avoid technologies that promote ideologically distorted or impoverished beliefs.

D.   Seek technologies that can enable disadvantaged individuals and groups to participate fully in social, economic, and political life. Avoid technologies that support illegitimately hierarchical power relations between groups, organizations, or polities

To help secure democratic self-governance: (Chapter 7)

E.   Keep potentially adverse consequences (e.g., environmental or social harms) within the boundaries of local political jurisdictions.

F.   Seek relative local economic self-reliance. Avoid technologies that promote dependency and loss of local autonomy.

G.   Seek technologies (including an architecture of public space) compatible with globally aware, egalitarian political decentralizaiton and federation.

To help perpetuate democratic social structures: (Chapter 8)

H.   Seek ecological sustainablility.

I.   Seek "local" technological flexibility and "global" technological pluralism.

FIGURE 5-1. A provisional system of design criteria for democratic technologies.

COMPLEMENTARY CRITERIA

This chapter and the preceding one develop criteria concerned with democratic community and work; succeeding chapters develop criteria concerned with diverse aspects of democratic politics. The nine criteria that result are summarized in Figure 5-1. The figure's categorization of criteria in terms of their bearing on democratic community, work, and politics is, however, somewhat artificial, for these criteria emerge from a single well-structured conception of democracy and thus tend, as noted earlier, to complement one another. This complementarity is what one would hope for and expect, given Chapter 3's suggestion that we regard democratic community, work, and politics as complementary principles governing all institutional settings.

One should not, however, overstate the degree of unity and complementarity among these criteria. There is also potential for conflict among them, especially when interpreting and applying them in specific contexts. Devising practical technological strategies that on balance satisfactorily advance democracy will thus occasion further democratic judgment and contestation.

Chapter 6

MACHINERIES OF POWER

"I am the Guardian of the Gates,
and since you demand to see the
Great Oz I must take you to his
palace. But first you must put on the
spectacles."

"Why?" asked Dorothy.

"Because if you did not wear
spectacles the brightness and glory
of the Emerald City would blind
you. Even those who live in the
City must wear spectacles night and
day. They are all locked on, for Oz
so ordered it when the City was first
built, and I have the only key that
will unlock them."

He opened the big box, and
Dorothy saw that it was filled with
spectacles of every size and shape.
All of them had green glasses in
them.

—L. Frank Baum, *The Wizard of Oz*

TECHNOLOGY AND IDEOLOGY

One necessary condition of democracy is that citizens have extensive
opportunities for developing insight into their social circumstances (see
Chapter 3). A central threat to such insight is "ideology," defined here
as beliefs that help perpetuate injustice by systematically distorting
people's understanding.[1] Ideological beliefs may, for instance, camou-
flage an illegitimate social hierarchy or impoverish peoples' awareness
of opportunities for self-development.

Chapter 2 showed that technologies, as an aspect of their polypotency, express or embody ideas and help constitute systems of nonverbal

communication and belief. Could technologies have the capacity to propagate, via latent nonfocal means, beliefs that are ideological? If so, it would be important to learn to scrutinize technologies and techno-logical orders in terms of their potential to function ideologically, to publicly debate and challenge those technologically communicated beliefs that seem ideological, and if necessary, to seek democratically to transform technologies that represent a serious source of ideological obfuscation.

The most important qualification concerning this argument is that the language of ideology-critique can be dangerously inflammatory. It is therefore important to bear in mind the contestability of the examples and theory that follow; they are intended as food for contemplation and debate, not as dogma.

Second, when particular technologies do seem responsible for propagating ideology, the appropriate practical response will vary de-pending on the context. Sometimes it will suffice to criticize a belief publicly, without having to do anything else to modify the practices or artifacts that previously disseminated it. At other times that solution will be insufficient, but an adequate remedy may be found by imple-menting other procedural democratic reforms or Part II's other design criteria. Thus, in these instances, a concern with ideology provides further support for actions warranted in any case. Finally, there may be residual instances in which technologically mediated ideology provides reason to alter or abandon a technology or design style that would otherwise be of little democratic concern. In short, it is democratically vital that citizens learn to:

> Debate and challenge ideological beliefs and modify, replace, or avoid technologies that prove democratically incurable of their propensity to inculcate ideologically distorted or impoverished be-liefs.

Let's call this Design Criterion C (see Figure 5-1).

Previous Theories

One can begin fleshing out this criterion by contrasting it with prominent existing theories of technology and ideology. One dimen-sion of technology's ideological potential has been delineated by theorists such as Heidegger, Marcuse, Habermas, and Ellul, who have undertaken a sustained critique of the exaggerated role of instrumen-tal rationality within modern thought and politics.[2] Such theorists argue, in effect, that a mode of thought they consider essential to contemporary technological practice—that is, a preoccupation with efficiently choosing technical means to accomplish ends that are arbitrary, obscure, or of dubious merit—has become pervasive. This has all but extinguished normative or other noninstrumental ways of thinking, speaking, and acting, and thus established subtle forms of enslavement.

For our immediate purposes, general critiques of instrumental rationality have no obvious implications for technological design or practice. Neither Heidegger, Marcuse, nor Habermas proposes, for example, restructuring technological practice or developing alternative technological styles. Habermas argues persuasively that one should relegate instrumental discourse to a subordinate position within the broader context of rejuvenated moral and political deliberation. But technological practice can evidently continue as usual so long as instrumental reason does not spill over and dominate all of social and political life (and provided that unjust capitalist relations of production are superseded).[3] He imagines that it is possible, as it were, to get technology out of politics (whereas I believe that, because technology is essentially a form of politics, one must get democracy into technology). Habermas and related critical theorists of technology do not consider such questions as whether certain technologies might not themselves engender pathological dependence on instrumental reasoning or whether there might be other ideological effects specific to particular technologies, rather than to technology as a whole.

An alternative perspective involves ideological views about technology that are not conveyed through technological artifice or practice. The beliefs that technologies are natural rather than socially contingent, that they are morally neutral, or that they fully determine social experience each qualifies as an instance, and various scholars have criticized such beliefs.[4] These are false views about technology that tend to hinder establishing a more democratic social or technological order. But most critics of these beliefs have not considered whether technology might function as one of their sources.

Thus we have noted theories concerned with ideology that (1) *originates* from technology (considered as an aggregate) and that tends to distort social understanding or politics generally, or (2) that is *about* technology but that propagates independently of technology. A third possibility of more concern here is that particular technologies or design styles (as opposed to technology in the aggregate) may become a source of diverse ideological views about themselves, technology generally, or other aspects of the world.[5]

Technologies as a Source of Ideology

Consider several examples of technologically mediated ideology, beginning with beliefs about technologies themselves.

One of the most serious and prevalent misconceptions about technologies is that they are natural or inevitable rather than the result of contingent social choices. This idea is dangerously false, because it hampers the establishment of a strongly democratic politics of technology. Where did it come from? Partly from contemporary technological artifice, practice, and style. For instance, many contemporary technologies are designed so that they deliver their focal results in experiential abstraction from the countless other sociotechnical structures and processes required to bring them about.[6]

A second factor is that the technical organization of work in industrial societies—including the traditional sharp separations between those engaged in research and design, those who produce, and those who distribute or sell—helps convey a misleading sense that today's technologies could not rationally have been otherwise.[7] This false belief is reinforced by the fact that most design activity occurs within organizations that largely conceal their internal processes from outside observation.

A third factor is that contemporary design procedures tend to limit even engineers' and designers' own awareness of the extent of technological contingency. A technology's designers cannot, for instance, know—especially under contemporary regimes of trade and military secrecy—about all the alternative designs that other people may have considered but for one reason or another rejected. Moreover, designers may know about some of the range of choice within their own domain of specialized experience, but they do not necessarily generalize from that domain to other domains. Finally, modern technological enterprise can involve huge teams of designers, engineers, and scientists working within a highly subdivided, compartmentalized, and increasingly mechanized division of labor, such that few participants have a comprehensively detailed overview of the entire process in which they are engaged.[8]

A fourth factor is that powerful actors (e.g., states and large corporations), once committed to a particular technology, tend to develop an interest in suppressing alternatives and even the social awareness of such alternatives. Government bureaucracies may act in this way because it enhances their ability to publicly justify their decisions as being technically necessary rather than value-laden and politically motivated.[9] Similarly, firms that have invested heavily in a particular technology sometimes stifle competition by patenting alternative technologies or by purchasing exclusive patent rights in order to prevent their exploitation.[10]

A fifth factor—partly a reinforcing historical outgrowth of the others—is that we lack a societal custom of subjecting technologies to critical democratic scrutiny. In the absence of a democratic technological politics, it has become customary to let combinations of distant bureaucracies or depersonalized market forces introduce technological systems that no single individual, group, or organization consciously chose or governs. This supports the false inference that because no particular person or group chose, the result is natural rather than a partly explicit, partly tacit social product.

A sixth factor is that—for complex historical reasons, often involving coercive technological influences or overt imperialism—there has been a global trend toward homogenizing technological style, a trend that tends to obliterate alternatives that would palpably demonstrate that present technologies are socially contingent.[11] This trend has been exacerbated by the prevalent ideological conviction that change is inevitably "progress," which contributes to ignoring, or dismissing as backward, local cultural and technological alternatives that manage to persist despite powerful contrary social forces.[12]

Thus the widespread misconceptions that technologies are not social products and that societies therefore have no democratically meaningful technological choices are at least in part a product of the technological order itself.

Can anything be done? Robert Schrank has described his experience in the 1970s trying to discuss alternative work arrangements with a group of General Motors Corporation union stewards and committeemen. First he asked them to imagine that they could arrange their work and their factory however they wished. How would they go about it? The initial answers seemed frustratingly orthodox: shorten the work week, reduce noise, fire some overbearing foremen, and so on. Trying to loosen up their imaginative energies, Schrank drew them a sketch of the innovative Volvo auto plant at Kalmar, Sweden. After diagramming the hexagonal plant geometry, the use of dollies rather than an assembly line, and a semiautonomous work team with no foreman, he asked, "Now, what about doing some things like this?"[13] Despite the fact that Kalmar was up and running successfully, the American auto workers reacted skeptically to the Volvo plant's innovations:

> That's ridiculous. I mean the company [General Motors] knows more about this stuff than we do, and if this was a good way of doing things, wouldn't they do it that way? . . . This company is in business to make money, not to run Mickey Mouse programs.

Writing afterward, Schrank commented:

> The frame of reference of these workers was the linear assembly line as they experienced it. Even to think beyond that seemed difficult.
>
> Based on our individual experience, we have little or no way of learning what the notion of a participative organization is about. All of our learning about institutions . . . , all institutional life—assumes a hierarchical order of things. After this lifetime of preparation, the auto workers I met . . . , like the rest of us, arrive at the workplaces prepared for a hierarchy and nothing else. So we settle into our positions on the pyramid and let the person above do the worrying.[14]

This example illustrates several themes. First, these workers' experiences of the technology of production within their own workplaces helped them conclude that the present design was the best or only one possible. Thus the technology's pervasiveness and stylistic homogeneity contributed to masking its own contingency. Second, Schrank surmised that a contributing factor was the workers' previous experiences at home, school, and elsewhere—settings that are also technologically shaped. Third, in addition to "learning" that technologies afford little opportunity for social choice, these workers had been technologically and experientially instructed to accept the inevitability of social hierarchy. Fourth, each worker's experiences in a variety of technologically shaped microsettings contributed to a generalized belief that not only were the particular technologies and social settings that they had personally known inevitable, but that technology in general and perhaps the social order itself were inevitable. Thus there was an ideological spillover from the social microlevel to the social macrolevel.[15]

Yet there is a sequel to Schrank's story. After the workers' initial skeptical and self-defeated reactions, and after a few more beers, Schrank suggested breaking into small groups, each of which would try to devise an original way to assemble automobiles. They did. Suggestions ranged from competitive chaos (dump the parts for an automobile in a pile and have work teams compete to see which could build a car faster) to variations on the Kalmar electronic dollies and also a modified assembly line in which teams worked in bays along the line, reducing the division of labor and linear work sequence. Although there was nothing shockingly original in the workers' proposals, they had taken a great step beyond their initial fatalism.

Other similar experiences suggest that when there is follow-through—when people actually begin to struggle together and realize structural social and technological change—there is a reverse anti-ideological spillover from the microlevel to the macrolevel.[16] After experiencing collective self-improvement and structural transformation at the microlevel, people can develop a sense of shared empowerment, a feeling of political efficacy, and an emergent realization that there may be other aspects of their world that they have been taking too much for granted. Even the limited achievements recorded by Schrank are sufficient to show that ideology is an adverse influence, not an irreversible determinant. With the help of Schrank's prodding, in only a few hours the ideology of technological and hierarchic inevitability dissipated among the auto workers, even if only partially and temporarily.

Finally, part of what made it possible to lift the ideological veil was Schrank's knowledge of a realized alternative (Kalmar). That reveals some additional limitations of Schrank's case: Kalmar was the only serious alternative of which Schrank's group was aware. Given also that time was short and the participants were not a large or experientially diverse group, it is perhaps not surprising that the alternatives they invented were bounded conceptually by chaos, the familiar, and the one known alternative. But that is sufficient to suggest that many recommendations advanced earlier on other grounds, as well as others still to come, can contribute to overcoming technologically mediated ideology.

For instance, job rotation, skill diversification, flexible careers, and citizen sabbaticals—especially within a context of cultural and technological pluralism[17]—can each be expected to contribute to expanding awareness of technological alternatives, to increasing political empowerment, and to enlarging creative capacities. These results would provide one basis for people inferring the potential malleability of technological designs and practices. Establishing more examples of communitarian/cooperative technologies would reinforce these results, helping particularly to unmask the apparent inevitability of hierarchy. If progress were also made toward democratizing technological design processes and publicizing their results, one could expect further lessening of ideology.[18] Finally, gradually creating a general democratic politics of technology, including contesting democratic design criteria, would provide a means of sharing and consolidating local gains in knowledge and becoming more democratically effective in dealing with translocal structures.

Public criticism of ideology and its structural sources could require supplementation by actual changes in technological design. For instance, modern technologies tend, as noted earlier, experientially to separate focal results from many of their necessary conditions and nonfocal consequences. Other democratic design criteria address this issue indirectly,[19] and flexible career patterns would also tend to counter its ideological effects. However, it remains possible that some technologies ought, even in the absence of further democratic grounds, to be redesigned to counter their tendency to obscure their own nature or other aspects of the world.

The specific beliefs we have examined, by masking social contingency, are of particular democratic importance. But the potential modes of technologically communicating ideology, as well as the variety of ideological messages communicated, are unlimited. Each example of technologically mediated ideology raises the possibility of requiring a somewhat different coping strategy. However, there is reason to expect that the strategies already considered—proceeding with democratization, publicly challenging ideological beliefs, and, if necessary, redesigning artifacts and practices to reduce their ideological propensities—is generalizable. This is because although each technology, as it appears in its own context, "tells" its own story, one can generally expect that technologies that are otherwise democratic will tell stories consistent with democracy, just as the converse will tend to hold true for nondemocratic technologies.

The following two examples demonstrate ways in which technologies can propagate ideological conceptions about aspects of the world other than technology.[20] Architecture is the technological medium perhaps most often recognized as harboring the potential to communicate ideas nonfocally, including ideas potentially subversive of democracy. Contemporary university architecture, for instance, ensconces each academic discipline in a separate building (or corridor or set of buildings). Groups of "related" disciplines' buildings tend to be clustered near one another and away from others (e.g., physics near chemistry, with both of these across campus from history and religion). This physically hampers creative interdisciplinary collaboration and the pursuit of transdisciplinary knowledge.

There are also more subtle architectural effects. University architecture does not merely hinder physical collaboration. It makes it difficult even to conceive of transdisciplinary knowledge. As Harari and Bell observe, "We have thus complemented conceptual categories and exclusions with physical and architectural configurations that mirror and reinforce divisions: walls, partitions, separate university faculties and libraries."[21] The very solidity of the buildings tends to stamp the knowledge that universities generate as enduringly valid and socially privileged. The generic structure of knowledge becomes taken for granted as rigid, compartmentalizable, and finite. The notion that knowledge and the world itself might instead be open-textured, flexible, ambiguous, nonfinite, perhaps even sacred, all but vanishes.[22]

Transdisciplinary and nondisciplinary knowledges have no comparably persuasive, materially embodied, authenticating symbolism. The vast majority of what people know is won through their own experience,

interpersonal communication (verbal and nonverbal), and critical reflection on those experiences. That knowledge can neither be codified
nor systematized.[23] Yet when confronted with architecturally authenti-
cated university knowledge, who has the fortitude to retain faith in his
or her own powers of discovery? (Perhaps people who become poets and
drive taxis.)

There are also nonarchitectural examples of technologies influencing how people understand and experience the world. For instance,
some technologies—such as telephones, television, hearing aids, windows, eyeglasses, public address systems, and computer networks—directly mediate human interaction and communication. In each case, an
individual's experience of the world is qualitatively altered—among
other ways, by sensorially filtering experience. A telephone, for example, reduces people to a "monodimension," a disembodied voice.[24] A
color television renders the infinitely faceted, encompassing quality of
live human experience as a rectangularly circumscribed, two-dimen-
sional image, with altered colors and sound, and no taste, smell, or
tactile sensory realms to experience.

Even nascent or hypothetical new electronic media that convey a
dimensionally richer sensory display are not a substitute for face-to-face
interaction, because electronic media implicitly choose how to decom-
pose holistic experience into analytically distinct sensory dimensions
and then transmit the latter.[25] At the receiving end, people can resyn-
thesize the resulting parts into a coherent experience, but the new whole
is invariably different, and in some fundamental sense less than the
original. In terms of Michael Polanyi's philosophy, part of what is lost is
that the original whole was partially constituted by a context that was
essentially tacit, open-textured, and nonspecifiable.[26] Hence, when one
analytically or technically decomposes a whole into parts, some of the
context that was essential to the original whole is invariably omitted.

Chapter 5 mentions such phenomena in conjunction with their
psychodevelopmental implications. The different concern raised here
is with the potential hegemony of such experience. Television advertising notwithstanding, watching something on a screen is vastly different
from actually being there. Contrary to a former AT&T promotional
campaign, long-distance telephoning is vastly different from "reaching
out to touch someone." Can concerted social efforts, or else design
changes, help people remember that technologically mediated interactions are qualitatively altered and sensorially deprived? Might democratic oversight be needed to help ensure a judicious balance between
the proportion of human interaction that is face to face versus the
proportion that is technologically mediated? If not, can we escape
seduction into experientially impoverished, easily manipulated, electronic worlds?

"Textbooks" for Democracy

Technologies can pose ideological threats. However, if approached critically—that is, with a concern for democracy, a sensitivity to polypotency, and some familiarity with examples of relatively democratic technologies—perhaps technologies can also be interpreted as "texts" to help identify undemocratic features of a technological or social order.[27] Thus, if some technologies obscure their own nature or other aspects of the world, other technologies may inadvertently reveal aspects of a society that are otherwise obscured. William Irwin Thompson, onetime professor of humanities at the Massachusetts Institute of Technology (M.I.T.), exhibited such a strategy when he used architecture to interpret the university's educational, engineering, and social philosophies:

> What distinguished M.I.T. from any other university was not its science but its overwhelming lust for power. I should have guessed earlier from the buildings, because M.I.T. generously gives itself away in its style of architecture. The conscious monumentality of the Great Court, the fortress shape of the political science building, the Persepolis staircase of the Student Center, and the colossus of the Earth Sciences Building, all declare the power of an institution that can locally transform the economy of Massachusetts, nationally contribute to the military and economic supremacy of the United States, and universally send rockets into outer space.[28]

For the purposes at hand, it is irrelevant whether Thompson's interpretation is "correct." Any person's interpretations are likely to intermix one-sided or arbitrary personal views with others that could win social validation. Only through an iterated, collective interpretive process can one discover what other people think and whether one's own considered interpretations coincide with theirs.[29] However, citizens will better understand their societies—a necessary step toward developing sound democratic judgment and embedding it in action—when this type of technological interpretation becomes integral to political practice.

POWER RELATIONS

Another fundamental democratic requirement is relative equality among citizens to shape the basic circumstances of their lives. Toward this end Chapter 3 proposes societywide organizing principles for achieving a system of democratic power relations. Technologies can influence or help constitute the structure of social power relations in many ways.

All of Part II's design criteria are implicated in the effort to democratize power relations. Criterion A prescribes relatively egalitarian power relations within communities, workplaces, and other institutions. Criterion B prescribes egalitarian opportunities for individuals to develop skills needed for effective political engagement. Criterion C highlights symbolic means through which technologies can help perpetuate hierarchic power relations. This section addresses two further dimensions of democratic power relations involving, respectively, comparatively disadvantaged and hyperadvantaged social groups.[30]

Equal Opportunity

All contemporary societies include groups whose opportunities for effective participation in social, economic, and political life are circumscribed. Identifying such groups is a contested proposition because power and social rewards are at stake. But it seems fair to suggest that among the disadvantaged in the United States are women relative to men, people of color relative to whites, those perceived as physically or mentally disabled, the unemployed, the poor, children, sexual nonconformists, non-English-speakers, and so on. Do technologies play a role in constituting disadvantaged groups? Let us examine several examples.

Women

Women have generally played a secondary role in political, economic, and technological decision making in industrial societies. The consequences are stamped in the technologies and architecture with which women must cope every day. "Labor-saving" appliances "liberate" many wives to do housework that was once performed by other family members: during the bygone era of open-hearth cooking, for example, men chopped wood and children hauled water, thus contributing more equally to household maintenance. Likewise, modern household and neighborhood designs often promote social isolation as well as a heightened risk of physical abuse, while restricting opportunities to organize childcare. Most public transit systems have been designed without regard to women's typical social responsibilities. How is a mother supposed to get a baby carriage up onto a traditional bus or down the steps of a New York City subway station? Many workplaces have jobs stereotyped as female that carry special risks of isolation, domination, stress, or health hazards. For example, secretaries and keypunch operators, who are predominantly female, suffer unusually high levels of stress-related emotional and physical disorders. Finally, reproductive technologies tend to subordinate women to a male-dominated medical establishment, scientific research programs purport to establish politically significant biological disparities between women and men, and mass media marketing techniques are frequently degrading to women or erect punishingly unattainable beauty standards.[31]

These consequences, each adverse in its own way, also conspire to limit women's opportunities to participate on equal terms in social and political life, including in the politics of technology. To explain these results, one need not invoke misogyny or conspiracy theory (although the temptation to do so may be strong). Generally, it seems more plausible to blame the indifference and insensitivity concomitant to male-dominated institutional hierarchies and design professions, in which women's evaluations of their own needs rarely figure as even a discussion topic.

Children

Colin Ward has documented innumerable ways in which contemporary urban design—generally oriented toward the lives of working adults— often stifles children's creativity, restricting their chances to develop self-esteem, efficacy, and awareness of the socially constructed nature of their world. Even spaces designed with children in mind can unwittingly produce these effects. Ward argues that "park and playground designers . . . usurp the creative capabilities of the very children who are intended to use their work by building play sculptures instead of providing materials for children to make their own."[32] In acceding to constructed environments that deny children an adequate role in shaping their world, there is a real risk of acceding to a population of future adults ill-prepared to contribute to democratizing their sociotechnological order.[33]

People with Disabilities

People with perceived mental or physical disabilities routinely suffer, on top of the pain that can be intrinsic to disability, societal unreadiness to accept them as moral agents or equals. They have frequently been isolated from society and one another, under conditions in which the extent of suspension of civil rights and human courtesies bears little relation to their degree of functional impairment. Technology has played a role here. Hazardous technologies are, of course, one cause of disabling accidents and illnesses. Diagnostic technologies are used to define standards of normality—standards cloaked often in a misleading language of objectivity and precision—and then to stigmatize those who do not measure up.[34] Disabled people who escape institutionalization often confront a built environment that exacerbates the consequences of disability by erecting obstacles to locomotion, communication, and ordinary social engagement.[35]

Technologies thus contribute to constituting certain groups as disabled and disempowered.[36] On the other hand, can disempowered groups, including the technologically disempowered, ever achieve reempowering technological transformations? Can technologies some-times play a fruitful role within struggles toward more democratic power relations?

Telecommunications and Empowerment

There are many cases in which isolated or marginalized groups have harnessed telecommunications technologies to self-organize or to assert themselves politically. For instance, during the American civil rights movement of the 1950s and 1960s, black leaders cleverly ensured that dramatic confrontations with white segregationists occurred in full view of reporters from the mass media[37]. More recently, dissident political movements around the world have mobilized mass uprisings against totalitarian governments with the help of telephones, photocopiers, videocassette recorders, and fax machines.[38]

Barrier-Free Design

During the past several decades, people with physical disabilities have organized themselves to press the case that buildings, streets, and trans-portation systems—and thus social, economic, and political life gener-ally—should be physically accessible to all citizens. Tangible results are evident in laws and building codes mandating wheelchair-negotiable buses, curbs, bathrooms, public buildings, and plazas; braille-encoded elevator buttons; public phones with adjustable volume; and so forth. Integral to this movement are claims concerning fundamental civil rights and also the lesson that disability—rather than being intrinsic—is defined and established relative to a socially constructed environment.[39]

Based on the preceding theory and cases, I propose as the first of two components of Design Criterion D that democratic societies:

> Seek technologies and a technological order that can help enable disadvantaged individuals, groups, and associations to participate

fully in social, economic, and political life. Avoid technologies that exacerbate social inequities or that perpetuate new inequities.

In a sense this criterion is already implicit in Criteria A's and B's prescriptions concerning egalitarian democratic community and self-actualization. However, it is crucial to be explicit, because the members of a disadvantaged group are often socially silenced or geographically dispersed among many communities in ways that make it easy and convenient to overlook their rights and needs.[40]

Suppose that addressing every legitimate claim of every disadvantaged group would catastrophically bankrupt an economy. Are there no practical limits to moral claims? One plausible response would be to redress social disadvantages until this begins to cause other commensurate harms to the basic democratic structure of a society (e.g., by commensurately disadvantaging others). A more explicitly neo-Rawlsian standard would be to redress social disadvantages until this begins to render the least advantaged worse off than they would otherwise be.[41] (Here strong democrats might wish to fortify Rawls by insisting that ordinarily the least advantaged themselves ought to play an important role in judging their own circumstances under competing policy scenarios.)

It is unlikely that the contest among competing standards of social redress will soon find a definitive resolution.[42] Even if there were a consensus on a single standard, applying it would occasion dispute. But if it is difficult to agree whether or when to stop redressing inequities at some hypothetical future time, in practical terms few societies are anywhere near the point at which this question need become paramount. Virtually all modern societies harbor sufficient resources to make significant headway—without fear of aggregate economic duress—in empowering their disadvantaged members. Thus in most instances the obstacles to empowerment have more to do with the concern of privileged groups or institutions to preserve undemocratic structures. This leads to the second component of Criterion D.

Hyperempowerment

Criteria A and B establish the need for egalitarian power relations and opportunities within communities and organizations. But what of the technologically mediated power of unduly privileged groups, organizations, or polities to influence external social circumstances? In other words, if democratic equality requires reempowering the disadvantaged, might it not also require erecting complementary obstacles to hyperempowerment? I thus propose as the second component of Criterion D that societies:

> Avoid technologies that support illegitimately hierarchical power relations between groups, organizations, or polities.

Examples follow of ways groups and institutions can, with technological assistance, circumvent the rough equality of power essential to democratic structuration.

Privileged Groups

Privileged social groups and classes have generally found the technological domain congenial for reproducing their social advantages. For instance, historically, the provision of city services (such as sewerage and public transportation) and the spatial distribution of housing classes relative to pollution sources have tended to reflect and help replicate disparities in wealth. Cultural systems of consumption—both mass consumption and conspicuous consumption—are also implicated in reproducing inequality.[43]

Organizational Size, Structure, and Power

The vast size, wealth, geographic command, hierarchic structure, and consequent external power of many modern bureaucracies (both corporate and governmental) are sometimes interpreted as necessary to cope with the size or complexity of the technological systems these organizations govern.[44] However, this argument assumes uncritically that the same technological hardware could not conceivably be developed and governed via alternative institutional means or that comparable focal ends could not be achieved via alternative technological hardware and systems.[45]

For instance, the San Francisco Bay Area's public transit services include trolley cars, several kinds of buses, light and heavy rail subway cars, high-speed ferries, government-sponsored van pools, and more. This very complex system functions well despite being designed and managed by more than 30 independent organizations without any overall system of centralized control.[46]

Even though it has not been well substantiated, the argument for a technology-based bureaucratic imperative nevertheless helps legitimate undemocratic organizational structures. For instance, prior to enactment of the Public Utility Regulatory Policies Act of 1978 (PURPA), U.S. electric companies used similar arguments to defend their regional monopolies against independent electricity producers. Yet the latter, once PURPA was implemented, began to contribute more to expanding U.S. generating capacity than did all the major electric companies combined.[47]

There is thus reason to doubt the unqualified claim that technological systems such as electric power grids, mass transit, petrochemical production, broadcast television, and urban wastewater treatment necessitate management by powerful undemocratic bureaucracies. A more plausible view may be that specific technologies supporting internal coordination and control are among the factors that make such organizational structures possible. For instance, the creation of large American corporations during the latter half of the 19th century depended on railroads and the invention of telegraphy, typewriters, carbon paper, mimeograph machines, and vertical filing systems.[48] The contemporary globalization of corporate and military operations likewise depends on sophisticated computer, telecommunication, aircraft, and aerospace technologies.[49]

Having evolved—with the help of technologies of internal coordination and control—structures that are hierarchic, resource-rich, and relatively unaccountable to external authority, large organizations are then able to deploy the same or additional technologies (as well as other resources) in ways that further enhance their power relative to citizens, polities, and other organizations lacking comparable resources. For instance, numerous analysts have found government agencies, private corporations, and political action committees acting much alike: they exploit their control of such things as research and development capabilities, biomedical screening and tests, advanced telecommunications, or mass media and computers to influence legislatures and electorates, to conduct domestic surveillance, or to pursue more subtle forms of social manipulation or intimidation.[50]

Conscious conspiracy need not always be operating to produce such effects. For instance, profit-seeking firms may try to replace communitarian technologies with individualized counterparts simply to enlarge markets for their products and services.[51] That the resulting community fragmentation also proves disempowering to citizens is merely a nonfocal side effect, albeit one useful for reproducing and expanding undemocratic power relations.

Another form of hyperempowerment may arise when an organization creates or governs technological systems on which many other institutions, people, or technological systems depend. It then tends to acquire potent ideological ammunition with which to disarm critics and help legitimate and reproduce itself. The usual form of the argument runs: (1) it is technically imperative to perpetuate the current technological system ("Lives and welfare depend upon it," "There is no alternative," etc.); and (2) the only way to perpetuate that system is to perpetuate the exact organization that currently governs it. I have already suggested fallacies in this argument's component elements, but there is no denying its ongoing political efficacy.

One follow-on strategy is that the same organizations that argue that "there is no technical alternative" sometimes take no chances, doing whatever they can to suppress potential alternatives (as the Renault automobile company did when it managed to block development of French electric cars in the 1970s).[52] This practice might aptly be labeled "elite Luddism."

On another front, entire governments—especially in impoverished developing nations—often feel thwarted or coerced by technologically empowered transnational corporations (TNCs). TNC critics tell of space satellites gathering strategic information concerning natural resources, cases of appalling environmental destruction and work conditions, biotechnology-based agribusiness threatening to displace entire sectors of third world agriculture, non-Western cultures overwhelmed by alien technological styles and media sources, and computerized telecommunications and transportation technologies permitting a fluidly adjustable global division of labor in which even small steps toward labor organizing or government intervention are parried swiftly with threats of plant closure and capital flight.[53] A few of these fears could be exaggerated. But predictable critics are not the only ones telling the tale. A 1990 *Business Week* cover story boasted of new "stateless" megacorporations "leaping boundaries" to finesse trade restrictions, intimidate labor unions, elude domestic political opposition, and "sidestep regulatory hurdles."[54]

Constraints and Possibilities

Technological hyperempowerment provides grounds for concern but not for despair. On the sobering side, those who envision modern technologies conveying societies effortlessly toward democratic utopia are dangerously naive.[55] Moreover, technologies alone are obviously not the whole problem. Technological power is thoroughly entwined with other sources of power such as great disparities in wealth, influential acquaintances, class stratification, disproportionate cultural influence, and so on. Thus, in strategic terms, to neglect any one power medium would be as naive and imprudent as neglecting another. Because power normally includes significant capability to reproduce itself, no one should expect change to result merely from articulating a proscriptive design criterion.

On the other hand, power is never absolute; powerful groups must contend with their own internal cleavages and contradictions. Power is also not comprehensive: if some technologies function partly as constituents or instruments of hierarchic power, others do not.[56] Moreover, structures, including technologies, invariably harbor ambiguities; recall disadvantaged groups that have found ways to exploit certain telecommunications technologies in their struggles for justice and self-determination.

One specific lesson from such stories seems to be that mere information—the commodity currently most hyped by high-tech-oriented government officials and corporations—has little political value to those not organized to interpret it critically and to act on it as they see fit. Access to computerized information can even prove harmful, especially if one takes into account the debilitating nonfocal effects often associated with its creation and distribution.[57] This is particularly true because hyperempowered organizations can sometimes use the same technologies and information sources to much greater relative advantage.[58] The disadvantaged groups that have made democratic headway with the help of telecommunications technologies seem to be those able to escape rapture with the prospect of copious quantities of information. Instead they find creative ways to use technologies as one vehicle for self-organizing and for coordination with their allies.[59]

There are also cases in which political or legal means have been exploited to help limit technological hyperempowerment. For instance, the 1971 defeat of a proposed U.S. government initiative to develop a fleet of commercial supersonic jets can be interpreted as a victory for environmentalism, middle-income taxpayers, and citizens generally over the interests of large aerospace companies (who would have built the expensive planes) and of corporate executives and the very wealthy (the only people who would have been able to afford the steep fares).[60]

In international affairs, developing and other nations' tactics in confronting powerful TNCs have ranged from capitulation to playing off TNCs or industrial states against one another, nurturing domestic industries able to compete in protected local markets or abroad, and reforming laws governing the chartering or behavior of corporations. These nations have also struggled to use multilateral forums and negotiations to establish more egalitarian international regimes governing global transportation networks, telecommunications, biotechnology, and common property resources (such as ocean fisheries and deep-seabed minerals).[61] Trade unions and public interest groups have demonstrated some effectiveness in assisting such efforts by organizing selective international consumer boycotts, exposing sales of hazardous products to developing countries, and collaborating politically with disenfranchised communities and indigenous tribal groups.

Sometimes a frontal assault on technological hyperempowerment may prove less fruitful than indirect approaches. For instance, the struggle for democratic community (Criterion A) would empower individuals within collectivities that would enhance their ability to resist external power intrusions as well as to press for further democratic reforms.[62] Implementing the criteria developed in the next chapter would enhance such empowerment. Moreover, efforts to democratize workplaces (as per Criteria A and B) would soften internal hierarchies and incorporate a broader range of social perspectives into decision making, thus tempering hyperempowerment from within. Tempering would also occur were democratic movements to impose limitations on state and corporate bureaucracies' size.

For these reasons, Criterion D's proposed restrictions on hyperempowerment may sometimes prove less significant as a near-term, practicable design criterion than as an analytic tool for assessing sources of structural resistance to other aspects of democratization, and hence for helping select important terrains of democratic struggle. Assuming that it is neither necessary nor feasible to challenge all undemocratic power arrangements at once, the ones to tackle first might be those that are weaker, that cause the most suffering or democratic harm, or that provide the greatest obstacle to other important transformational objectives. Criterion D may also help motivate inquiry into the degree of latent organizational flexibility associated with technological systems historically managed by undemocratic bureaucracies. Finally, it can play the same inspirational and empowering role as other democratic design criteria by underscoring the social contingency inherent in current sociotechnical arrangements and by providing compelling moral grounds for seeking more democratic alternatives. Should that sound unrealistic, one need only recall the effectiveness with which people with physical disabilities have used moral argument to win support for barrier-free designs.

We have considered democratic design criteria concerned with community (A), work and self-actualization (B), ideology (C), and power relations (D). The next three criteria each deal with technologies' influence on political processes, especially democratic self-governance.

Chapter 7

"WE WOULD CALL IT TREASON"

Technology and Self-Governance

Strong democracy calls for a relatively high degree of local self-governance, in part because an everyday citizen has greater potential to exert influence in smaller, immediate settings than elsewhere. However, technologies often affect people who are not directly involved in choosing or using them, including people dispersed over wide areas (see Chapter 2). A preliminary formulation of a democratic design criterion follows: Insofar as technologies that affect people in more than one community can promote a transfer of decision-making authority to translocal political arenas, it is desirable to avoid deploying technologies that unnecessarily occasion such democratically adverse jurisdictional shifts.

Technologies can affect a community's ability to govern itself in several ways. For instance, (1) a local technology can generate pollution or have other impacts that adversely affect other communities; (2) a local technology's operation can also render the community where it is located dependent on other communities or on institutions not located within the community; or (3) an integrated technological system can serve or connect more than one community. Each of these modes of influence can be formulated into a distinct democratic design criterion.

TRANSLOCAL HARMS

A local technology that harms people in other communities can provoke intercommunity conflict, in turn precipitating intervention by higher political authorities that subverts local self-governance. In the late 19th and early 20th century, American cities imported clean water or filtered and treated incoming water, while discharging raw sewage into rivers and lakes. Various methods of sewage treatment were known and others were under development, but few cities were willing to pay for them unless local sewage was causing local harm. Because the buildup of sewage increased illness and death in downstream communities, state governments passed preemptive laws protecting water quality, established state boards of health to administer the laws, and created new regional authorities ("special districts") charged with integrating and managing the systems of water supply and sewage treatment.

As the result of this state intervention, water quality and public health dramatically improved, but local autonomy dramatically declined. Worse, state-level action in the realm of public health set a precedent later implemented in developing institutions to govern the development of roads, ports, energy sources, and telephone service. The failure of municipal governments to assume technological responsibility for neighboring communities subverted their own autonomy and the tradition of local self-governance.[1]

Moreover, in the case of water management, it is evident that the trade-off between public health and democracy was unnecessary because, for decades after the establishment of regional authorities, the preferred—and quite successful—technologies remained chlorination or filtration of incoming water. These could have been implemented on a local basis (in contrast with upstream treatment of sewage prior to discharge, which indeed demands some form of transmunicipal coordination). Furthermore, even had states wished to encourage earlier adoption of predischarge sewage treatment, there were less-centralized institutional alternatives to establishing new regional authorities. For instance, state governments could have set and enforced minimum acceptable water quality standards or provided financial incentives for local compliance, leaving it to localities themselves to devise and administer their own technical solutions. Thus either way it would have been practicable to prevent disease without so severely compromising local civic authority.[2]

Today a related pattern continues to play out as large business corporations repeatedly use cross-border pollution as a rationale to justify environmental regulation at increasingly higher levels of political aggregation (i.e., shifting from local to state, state to national and, finally, national to international authority). When "successful," this reallocation of power has transposed environmental decision making to arenas relatively inaccessible to grassroots participation, where corporations have secured weak environmental standards that preempt the stronger standards favored at the local level. This logic helps to explain industry support for the 1970 U.S. Clean Air Act and for the 1990 amendments to the international Montreal Protocol (an international accord that regulates emissions of industrial chemicals hazardous to the earth's atmospheric ozone shield).[3]

Pollution that harms ecosystems or people's health is a classic example of a translocal harm. One can capture the concern that translocal effects could undermine democratic self-governance by suggesting:

> It is democratically vital to seek to restrict the distribution of potentially adverse consequences (e.g., environmental or social harms) to within the boundaries of local political jurisdictions.

In terms of Figure 5-1, this is Design Criterion E.

Economists call such translocal effects "negative externalities." However, here that term will encompass a broader range of phenomena than most economists take into account, including political, psychological, aesthetic, and other potentially adverse technological consequences.[4]

Reasonable Self-Restraint

Criterion E in effect asks that individuals, groups, and communities voluntarily exhibit a measure of self-restraint in selecting and governing technologies. But it is a self-restraint grounded both in the interest of others and, ultimately, themselves. Failure to exhibit such restraint would, in the first instance, abrogate our moral duty to show others the same consideration that we would wish them to show us.

However, suppose people knew that democratic institutions would ensure fair compensation for any harm that they caused, wittingly or not, to others. Why bother to exhibit self-restraint? One reason is that in some cases (e.g., when sacred values such as family heirlooms, ancestral community treasures, human lives or limbs, friendships, community, or democracy itself are at risk) adequate compensation may be impossible.[5] More generally, such thinking would progressively erode the possibility of local sovereignty and self-governance. True, in the short run individuals and communities may be tempted to impose adverse effects on other communities or simply fail to take these communities into account. However, the predictable long-run consequence—as in the case of 19th-century sewage management—is that those harmed by others' technological indifference will seek redress in higher level political forums, ultimately undermining local self-determination.

Although such self-restraint is often absent today, it is not unrealistic to expect it to be exhibited more frequently within a more democratic society. Most importantly, in a strong democracy rough equality of power would militate against being able to ignore those who today appear too powerless to win political redress. (It is also at least conceivable that future citizens who will have grown up and lived in more strongly democratic institutional settings may—thanks to enhanced opportunities for moral development—prove less likely than today's citizens to behave in narrowly self-interested ways. But it would be imprudent to bank heavily on that kind of general moral improvement.)

Even if some other communities chose to behave irresponsibly, a complying or injured community could, if necessary, expect to win the argument at the federal level. This would be true, first, on the grounds that offending communities are disregarding others' well-being (which violates the categorical imperative). Second, a harmed community could appeal to other disinterested communities on the grounds that the offenders' democratically irresponsible behavior—by provoking federation-level involvement—threatens the structure of decentralized federation itself. In short, it makes sense to argue it out at the federal level rather than to simply start behaving irresponsibly.

Even in the absence of an existing, strongly democratic social order, there are reasons to try now to establish a technological order more consistent with democratic self-governance. First, if technological systems are created that generate extensive translocal harms, this will tend to elicit the involvement of high-level political institutions, in turn denying local communities and citizens hands-on political experience useful for working toward democratization generally. Second, the mere existence of a technological order that generates extensive adverse externalities can constitute grounds for arguments against the possibility of democracy. Opponents can argue, in effect that "strong democracy would be wonderful, but it can't work given the current need for centralized high-level management of technologically induced social conflicts."[6]

Locally Concentrated Harms?

Why does Criterion E omit adverse external effects that remain concentrated within a local political jurisdiction? Because my concern here is with establishing a decentralized federation and local self-governance, and only translocally distributed externalities bear on that question. One could, of course, advance another design criterion governing intralocal harms, but that does not seem democratically necessary.

Assuming that local communities are egalitarian and self-governing, it should be up to them to decide for themselves—apart from satisfying other democratically necessary design criteria—which intralocal harms or risks they find acceptable.

However, if a local polity exhibits illegitimate hierarchy, there is a risk that disadvantaged citizens will suffer technologically imposed harms from more powerful neighbors, neighboring business facilities, or irresponsible local government. That would directly violate the categorical imperative and provide warrant for federation-level intervention. But the highest order grounds for intervention would be to try to remedy the inegalitarianism itself, of which the unfair local imposition of harm is a symptom. The legitimate exception would be as an interim, second-best measure—that is, to protect disempowered local citizens from imminent danger—while working on a high-priority basis toward local moral and political equality.

Implicit in these remarks is not callousness but a respect for local communities' relative autonomy. Democratic theory and practice must try to distinguish between adverse consequences that are voluntarily versus involuntarily imposed. Today public policy regarding technology tends to be preoccupied with managing a wide variety of technologically induced health, safety, and environmental risks. Strong democratic theory concurs that people must not be forced to run risks that others impose on them without their knowledge or uncoerced consent. But the paternalistic perpetuation of biological life is not a highest order democratic value. If people choose freely and knowingly to risk their own health or lives, perhaps in return for self-actualizing experiences not otherwise attainable, that is their privilege.

Contemporary preoccupation with health and environmental risks, narrowly defined, evinces a concern with the brute temporal duration of human life that becomes pathological if not tempered by a commensurate concern with the quality of life and with instituting democratic structuration.[7] Such one-sided fixation serves an ideological function to the extent that citizens, while busily worrying what governments and businesses are doing to protect their lives, are diverted from their deep interest in probing and striving to advance their society's structural degree of democracy.[8]

Externalities and Scale

In a sense Criterion E is an outgrowth of the frequent claim that a socially appropriate technology would be small in scale.[9] Yet Criterion E need not correlate tightly with other, more common measures of technological scale. A large industrial facility can be a prodigious generator of adverse externalities but so can thousands of air-polluting private automobiles or small wood stoves.

It is unorthodox to treat technological scale as significant especially in terms of the potential to export harm. Scale has more often been discussed in terms of the physical size of artifacts or the number of people engaged in a common technological endeavor. What of the democratic significance of other scale dimensions?

Authoritarian social relations may become more probable among large numbers of people if the need to coordinate complex interactions justifies establishing an organizational hierarchy with the accompanying risk of democratically illegitimate power asymmetries. On the other hand, although some proponents of appropriate technology have idealized small individualized technologies, communitarian/cooperative technologies would, by definition, have to be large enough to serve or be managed by team- to community-sized groups.

However, ultimately what matters democratically is not the number of people, but the quality of their interrelations and activities. Introducing "number of people" as an independent design criterion could thus prove overrestrictive (to the extent the correlation with conviviality, self-actualization, or hyperempowerment is not tight), would largely be redundant (e.g., given the consideration of scale already incorporated within the discussions of Criteria A and D), and could easily divert attention from other sociotechnological attributes bearing on each criterion. There are, for example, many workplaces smaller than Boimondau that are nevertheless much more authoritarian.

If the number of people engaged in a common technological endeavor matters, but only matters indirectly, what about the number of people or communities served by a common technological system? If 2, 10, or 1,000 communities are served by a single electric power station or dry cleaning establishment, should committed democrats be perturbed?

Consider the case of a large electric power station. A substantial amount of the public concern engendered by such plants arises because all of them do, in fact, increase the possibility of significant, widely distributed, adverse involuntary consequences. For instance, oil-burning electric plants can render communities dependent on the politically unstable Middle East, enhancing the risk of international war. Today's nuclear plants pose a risk of catastrophic releases of radioactivity and create nuclear weapons-usable plutonium as a by-product. But if someone were to invent an electricity-generating station that, say, ran reliably and cheaply on ordinary nitrogen (which is superabundant in the atmosphere) and produced no dangerous by-products, would that not diffuse much public concern with electricity production? Thus Criterion E already captures much of the well-founded concern with broad-based dependence on large technological facilities.

However, there are residual reasons for concern about broad-based dependence on a common technological system. One is that it may be difficult to achieve a consensus among the large number of people potentially affected on whether or not such a system is democratically adverse. This creates some presumption against a technology being large (in the present sense of serving many communities) but probably ought not absolutely to preclude all such technologies (e.g., in instances when an informed democratic consensus favors one, perhaps because it promises to facilitate compliance with other democratic design criteria). This and related considerations are incorporated in Design Criteria F and G. The preceding chapter has, moreover, already addressed the concern that those who control such systems can become collectively more powerful politically than the communities they serve.

Finally, when one thinks of technological scale, perhaps the dimension that first comes to mind is technological artifacts' sheer physical size. There may be aesthetic reasons object to or favor large-sized artifacts. Perhaps many people feel belittled by monumental artifacts or, alternatively, uplifted by the thought that these are human constructions. Some people prefer an uncluttered view of forested mountains to one obstructed by skyscrapers. But these are matters that, to the extent that they are not already reflected in Criterion E, ought to be handled through ordinary democratic politics. They do not fundamentally affect democracy's possibility.

Physically large artifacts and systems may, however, also raise questions regarding the ease with which they can be altered or abandoned if they should subsequently prove antidemocratic, or if citizens decide on other grounds that they do not want them. Chapter 8 addresses this issue.

Appropriate Technology as Democratic Technology

Harvard professor Harvey Brooks has argued against appropriate technologies on the grounds that small-scale technologies could only be widely deployed if they were produced, marketed, and serviced by large, hierarchically organized firms. Thus a necessary condition for appropriate technology would, reasons Brooks, negate one of the ends (convivial work) for the sake of which appropriate technology was initially advocated.[10] Is he correct?

There is something to Brooks's claim, but as a decisive critique it seems in one respect overstated and in another respect misconceived. The overstatement lies in the basic claim that nonauthoritarianly produced small-scale technologies would necessarily be too expensive to become widely deployed. As Brooks himself acknowledges, less authoritarian workplaces often prove more efficient than their rigidly organized counterparts.[11] Nor is it always true that large, vertically integrated organizations can capture vast economies of scale in production unavailable to smaller firms.[12] Thus many small-scale technologies could undoubtedly be produced at competitive cost by relatively small, nonauthoritarian firms. Even if this were not the case, small-scale technologies could nonetheless be democratically produced at a cost penalty socially warranted to the extent that it does not translate into other, offsetting structural democratic impairments. Finally, even if in certain instances both of these counterarguments were inapplicable, wouldn't it be better to produce appropriate technologies authoritarianly than, as is often the case today, to produce inappropriate technologies authoritarianly?

Nonetheless, it is at least conceivable that there are some small-scale technologies that could only be made widely available if produced and serviced authoritarianly or at a tremendous cost disadvantage. Thus, although Brooks argument is much overstated, it is not entirely specious. Moreover, his criticism is useful for highlighting a historical tendency among appropriate technology enthusiasts to assume that small-scale technologies will necessarily be produced by small-sized firms or that the question does not matter. Small-scale technologies can be produced nonauthoritarianly, but that is not inevitable, especially in the absence of direct political effort. For example, while proponents of appropriate technology have often expressed enthusiasm for the democratic potential of solar photovoltaic electric cells, by the mid-1980s multinational oil companies had purchased every significant solar photovoltaic company.[13]

Aside from being overstated, Brooks's argument is also misconceived, because it takes for granted the notion that appropriate technologies need to be small. Several lines of reasoning suggest that many appropriate technologies—here defined to mean democratic technologies—would be intermediate in size (neighborhood- or community-scaled). This begins to reveal weaknesses not only in Brooks's reasoning but also in that of many appropriate technology theorists. The latter tend to underemphasize or neglect both democracy's preeminent significance and its necessary conditions. This is manifested, for instance, in their failure to observe the democratically adverse nature of integrated systems of authoritarian, mass, and individualized technologies.[14]

Intermediate-sized technologies have the potential to optimize scale economies in pollution avoidance or abatement. They are also large enough to offer convivial social relations yet less prone than larger technologies to elicit an extensive vertical hierarchy to coordinate the activities of many people. Intermediate-sized technologies can nonetheless exhibit a measure of internal specialization in jobs, making it feasible for workers to undertake relatively complex and interesting tasks. Intermediate-sized workplaces and facilities can be more widely dispersed geographically than large ones, thereby lowering the cost of transporting and distributing their products. Finally, intermediate-sized technologies would not be produced in the large numbers that would make it economically attractive to mass-produce them. This could be desirable, because they could be design-adapted to local circumstances and provide more opportunities for local participation in their design, while contributing to an overall pattern of cultural and technological pluralism within the larger society.

In short, contrary to Brooks's claim, small-scale technologies need not in general be produced in an authoritarian fashion; nor is it necessary that socially appropriate (i.e., democratic) technologies be small.

Developing a conception of appropriateness more grounded in democratic theory would also counter Brooks's further claim that "[appropriate technology] and current technology are complementary rather than mutually exclusive."[15] If by "appropriate technology" one means democratic technologies, this argument is clearly false. Deploying democratic and undemocratic technologies side by side is often technically feasible, but it does not add up to a democratic sociotechnological order. Indeed, continued deployment of undemocratic technologies tends materially and ideologically to impair the prospects for deploying democratic alternatives.

ECONOMIC SELF-RELIANCE

The preceding sections explored the potential for translocal harms to provoke federation-level intervention. A second means by which technologies influence the possibility of local democratic self-governance is through their role in structuring intercommunity economic relations. Suppose, for example, that a community is dependent on technological systems controlled by other communities or by an organization located elsewhere. The first community's ability to influence those systems or the controlling organization is then limited—and even more so if many other communities are dependent on the same system. Alternatively, suppose a community or workplace produces goods or services that it sells in geographically extensive economic markets. In that case, it risks becoming dependent on faraway economic and social forces that, again, it can hardly influence. People may, as a result, be impelled to behave in ways contrary to their best democratic interest. In other words, technologies can help to establish patterns of economic interdependence that influence local communities' capacities to govern the fundamental conditions of their existence, including technologies themselves. This suggests that it is democratically vital to:

> Seek technologies that contribute to relative local economic self-reliance, while avoiding technologies that promote dependency and loss of local autonomy.

In terms of Figure 5-1, this is Design Criterion F.

Criterion F refers to *relative* local economic self-reliance, not *total* self-sufficiency. There is nothing democratically wrong with a community or business choosing to participate in transcommunity economic exchanges. However, communities should democratically monitor and, as warranted, govern the number, types, and aggregate structural implications of these exchanges to ensure that they do not undermine the possibility of local self-governance. If a community is presently far from economically self-reliant, political effort is required to plan and implement a transition.

A century ago, for example, London differed from other leading world cities in eschewing reliance on a single major electric company, large generating plants, or even a citywide electric grid. Instead, dozens of small electric companies scattered throughout London—some privately owned, some publicly owned—deployed a diverse array of small-scale electrical generating technologies. Was London backward and irrational, as engineers from elsewhere commonly supposed? Not obviously. London's electric companies operated at a profit and provided reliable power adapted to local needs. London's subsidiary borough governments, seeing their own political significance and autonomy as inextricably bound to the infrastructures on which they depended economically, consistently opposed parliamentary efforts to consolidate the grid. The boroughs favored a highly decentralized electrical system that each could control more easily.[16]

For analogous reasons, a number of American cities, towns, and neighborhoods have begun to encourage new, locally owned businesses oriented primarily toward production for nearby markets. The nonprofit Rocky Mountain Institute has developed an analytic process, along with supporting instructional materials, that a growing number of towns in economic difficulty are using to reduce their consumption of imported energy, water, and food (rather than to rely on distant supplies) and to reinvest local capital (rather than to put it in the hands of bankers or investment managers hundreds or thousands of miles away).[17] Once these communities are more secure against distant market forces or transnational corporate decisions, they are more empowered to conceive and undertake local democratic initiatives.[18] Local self-reliance can, for instance, gradually empower communities to resist the coercive threats of corporate capital flight.

The pursuit of self-reliance contrasts strikingly with local and state governments' prevalent strategy—self-defeating when it is not futile—of using concessionary tax breaks, waivers on environmental standards, or the promise of low wages to try to entice geographically fickle corporations. Similarly (and independent of which political party occupies the White House) contemporary U.S. trade and technology policies have remained preoccupied with advancing the nation's international competitiveness in ways that will assuredly erode local self-reliance.[19]

Communities that neglect self-reliance sometimes find themselves communities no longer. While London's borough governments were busy resisting creation of a citywide electric grid, in Germany and the United States many suburban and rural communities did not attend to developing their own municipal technological infrastructures. As a result, they found themselves politically annexed to nearby cities as a condition of obtaining externally provided clean water, sewerage, new roads, streetcar service, or electrification.[20] Thus they sacrificed not only their power to govern these new technological systems, but also the entirety of their independent political identity. Similarly, today cities and towns around the world have begun learning through harsh experience the repercussions of allowing their economies to be dominated by businesses headquartered elsewhere. Outsider-controlled firms have routinely shown themselves indifferent to the adverse local impacts of their decisions.[21]

In none of the preceding cases did technologies alone determine whether a community would become less or more self-reliant and self-governing. Electrification was not the only factor influencing the degree of self-determination exercised by London boroughs. However, in each case technologies played an important role.

To the extent that a community does participate in transcommunity economic exchanges, it is better if these are generally regional rather than global.[22] This makes it easier for a local community to monitor the structural significance of its economic interdependencies and to influence emerging developments of democratic concern. From a federation-level perspective, it is similarly helpful for regions to remain relatively self-reliant economically with respect to one another.[23]

(Third world elite rhetoric is rich in the language of self-reliance. This normally translates practically into the quest for greater national status or power in the current international world order. Unfortunately, leaders' pursuit of national self-reliance usually works to the detriment of local self-reliance.[24])

It seems prudent, in addition, that a community's translocal economic exchanges be limited, if possible, to relatively inessential goods and services, especially with respect to transregional trade and interdependency. That way, if it becomes democratically desirable to curtail a pattern of exchange, this can occur with relatively little trauma.[25] Modern Switzerland provides a partial example through its well-established contingency planning for total self-sufficiency in times of international crisis.[26]

To whatever extent a locale is producing for export to distant markets, it is still democratically preferable to do so under circumstances of local ownership or control and with extensive reliance on local skills and resources.[27] Such production remains democratically problematic because it means dependence on hard-to-foresee and uncontrollable market forces. But locally controlled production is partially buffered from vagaries in the supply of inputs and also from the vagaries of direct subordination to policy decisions made in distant corporate headquarters and national capitals.

Finally, even if one's community has the opportunity to participate in a pattern of economic exchange with politically and economically weaker trading partners—thus suggesting that one need not fear loss of one's own local autonomy—there are at least two good reasons not to do so: (1) the pattern of asymmetrical power could suddenly shift, leaving one's community vulnerable and unbeloved (witness the sudden emergence to international prominence of the Organization of Petroleum Exporting Countries [OPEC] during the 1970s); and (2) participating in a system of asymmetrical and exploitative power relations, regardless of whether or not one is among the apparent beneficiaries, violates the categorical imperative.[28]

If translocal economic exchanges are democratically so dangerous, why not simply rewrite Criterion F to suggest absolute local economic self-sufficiency? It isn't necessary and could prove democratically adverse. Translocal economic trade and interdependence can, for example, provide access to materials, goods, and services that are not locally available; that would be prohibitively costly to produce locally; or that can be acquired more economically through exchange. Such benefits are democratically acceptable if restricted in quality and quantity to remain consistent with local self-governance and with other necessary conditions of democracy.

Another reason a community might permit translocal economic exchange is to promote selective economic competition in the interest of detecting and reducing local economic or institutional inefficiencies. However, as always, it is crucial not to sacrifice democracy in favor of pure economic gains. That would be "efficient" only in the prevailing narrow and ideological meaning of the word.[29] For instance, imagine an innovation in workplace technology—such as introducing assembly-line methods into office work—that enhances productivity, while at the same time sharply reducing workers' opportunities for creativity and self-management. Whether the postulated productivity gain is socially good would depend partly on whether what is being produced is democratically benign. Either way, unregulated competitive pressure could compel local firms to adopt such a technology with unequivocally antidemocratic consequences for workers (e.g., in terms of Figure 5-1's Criteria A and B).

Other Benefits

Self-reliance has further democratic virtues. Greater self-reliance implies developing a more diversified local economy, which would—especially within the context of flexible work scheduling—afford greater possibilities for creative self-actualization. The fact that people would encounter fellow community members in a more diverse array of contexts is beneficial for perceiving one another holistically rather than as one-dimensional role performers. Economic diversification, in turn, means living in communities that would become more richly textured and that would respond more smoothly and resiliently to translocal forces and events.

Local self-reliance implies working within one's community for the community's own good, which should contribute to a greater sense of local commonality. It implies a greater preponderance of communitarian/cooperativism or of individualized production for local use. It means that people would more often see the results of their labor being enjoyed—whether by themselves, other identifiable people, or the community as a whole. This contributes to personal satisfaction, self-esteem, and mutual respect.[30] Good in itself, such self-esteem is also a constitutive element in personal moral growth. Local production, for local or nearby use, also implies the possibility of closer adaptation of goods and services to local circumstances.

For the individual, a more diversified, self-reliant economy—coupled with opportunities for flexibly scheduled, multiple careers—reduces the danger of ever becoming fully unemployed. That can benefit the individual but also strengthens community stability, by reducing citizens' economic compulsions to move in order to find work.

Greater local self-reliance promises benefits to democratic knowledge. People would more easily and naturally know the circumstances under which the goods and services they use were produced, and under which those they produce were being used by others. This is crucial to developing sound moral and political judgment. (Criterion E's prohibition against translocally distributed harms strongly complements this aspect of Criterion F.)

In contrast, Milton Friedman is famous for celebrating the anonymity of far-flung market exchanges and the consequent protection afforded against discrimination: "The purchaser of bread does not know whether it was made from wheat grown by a white man or a Negro, by a Christian or a Jew."[31] Perhaps so, but then again, neither does the purchaser know whether that grain was grown on farms that exploited migrant workers or, say, by a chemically intensive mechanized agribusiness that expropriated a dozen traditional farming communities, promoted topsoil erosion, contaminated rivers, and increased regional dependence on imported oil. Growing recognition of the importance of such knowledge is evident in contemporary consumer activists' efforts to require more informative labelling practices (e.g., tags identifying locally made products, organically grown foods, or cooperatively produced goods). Moreover, given that market economies have not infrequently been associated with deep and persistent racial discrimination—consider both U.S. and South African history—it seems that strong democrats must gird to combat discrimination head-on and through such structural measures as promoting citizen sabbaticals.[32] In short, market anonymity provides no sure fix for discrimination, but it does filter out politically vital knowledge and thus abet atrophy of moral capabilities.

Moreover, by being better able to witness and participate in the production of the local technological order, people might become less prone to misperceive technologies as natural or inevitable rather than as contingent social products. Local production for local use also implies closer experiential integration of technological means and ends,[33] encouraging a more comprehensive perception of technologies' full range of focal and nonfocal effects.

As noted earlier, adverse externalities are responsible for some popular opposition to large technological systems that serve many communities. However, from the standpoint of local self-reliance and self-governance, there is reason to be concerned with extensive dependence on such systems even in the absence of externalities. If a local community no longer controls or regulates one or two of such systems as water supply and treatment, energy supply, transportation, its own fire department, acute medical care, or food supply, perhaps that doesn't really matter. However, if it no longer controls or regulates many or all of those systems, citizens may recognize, in a general way, that local government no longer governs many important aspects of their community life. Once that perception arises, it can become a self-reinforcing prophecy, discouraging citizen involvement in local public affairs and in turn contributing to the further atrophy of local government capabilities.[34]

Many appropriate technologists have perceived a special causal link between small-scale, distributed energy technologies and political decentralization. Amory Lovins's work has often focused on energy because it is

> pervasive, symbolic, strategically central [and could] . . . as an integrating principle . . . be catalytic [for] . . . profound cultural transformation. . . . In an electrified society, everyone's lifestyle is shaped by the homogenizing infrastructure and economic incentives of the energy system.[35]

Such reasoning may be slightly off the mark. Insofar as energy is symbolically distinctive in American culture,[36] energy policy could conceivably play a catalytic role in social or cultural change. Energy also plays an essential material role in contemporary life and, episodically, a strategic role in politics. Yet if one is concerned with the relationship between technology and sociopolitical structure, it is not obvious that energy technology matters more than other kinds of technology (such as communications, transportation, architecture, water management, manufacturing, etc.).

Increased local self-reliance ought, by facilitating local government's capacity to govern local circumstances, to increase citizens' incentive to participate politically.[37] As local government becomes more potent and consequential, and as more citizens choose to become involved, local government's capacity to function as a mediating structure for popular democratic empowerment should also increase.

There is no guarantee that smaller, more locally controlled technological systems will be well managed. Like today's large systems, some would doubtless be managed well, some poorly. Either way, at least when dealing with more localized technological systems, citizens and groups can have more say in how systems are designed and operated. (This is especially true, of course, if localities are strongly democratic—which, completing the circle, becomes more possible and probable if they are relatively self-reliant.)

This establishes Criterion F's compatibility with conditions of democracy other than self-governance. (Conversely, movement toward implementing other democratic design criteria—e.g., devolution from organizational gigantism—would tend to facilitate progress toward self-reliance.) Such diverse, complementary democratic benefits provide grounds for promoting a somewhat greater degree of local economic self-reliance than might be warranted solely to facilitate democratic self-governance. On the other hand, because there is considerable ambiguity in how best to define and promote self-reliance, one can expect this criterion to remain vigorously contested.

Relative self-reliance promises to contribute importantly to strong democracy, but it is certainly not sufficient by itself to establish strong democracy. For instance, Nazi Germany and white-supremacist South Africa are both noted for having developed technologies to support national self-reliance—primarily to achieve the state autonomy required to pursue barbarously antidemocratic policies.[38] Even local self-reliance, if controlled by an unaccountable local power elite, can hardly be judged democratic. Democratic design criteria must work together as a complementary system within the broader context of democratic structuration.

Many appropriate technology theorists have advocated economic self-reliance but, because they fail to emphasize democratic reasons, risk legitimating antidemocratic variants. Many have also failed to discriminate between purely individualistic (e.g., household) versus community-level self-reliance or between relative versus total self-reliance.[39] Meanwhile, recent advocates of national technology policy—even the minority of advocates who are sensitive to democracy—overlook arguments in favor of partial self-reliance, taking it for granted that production must be oriented toward national and world markets.[40] (The extant degree of national integration into world markets does presumably make some attention to international economic competitiveness democratically important in the short run. But a concern to advance democratic self-governance would seem to concurrently mandate a gradual deliberate shift toward greater regional and local self-reliance.)

Because few modern communities have made even small efforts toward achieving self-reliance, the technical and economic possibilities for improvement are considerable, even taking into account only those technologies that presently exist. Furthermore, if one considers new technologies that might be developed were there a deliberate effort to look for them, then the possibilities for achieving greater local self-reliance seem great indeed.[41]

For example, today access to raw materials that are scarce or geographically not evenly distributed appears to be diminishing in economic importance; this expands the potential for self-reliance.[42] Similarly, flexible manufacturing technologies that facilitate economically producing small batches of products increase the potential for diversifying local productive capabilities, as may new telecommunications networks (which could allow remote communities to benefit from a range of problem-solving expertise that would be uneconomic to maintain locally).

Provided that they are found democratically acceptable on other grounds, new biotechnologies, too, harbor potential to help advance local self-reliance—for example, by creating new, renewable feed stocks to replace locally scarce industrial inputs or by permitting a much wider range of food crops to be grown under local ecological conditions. On the other hand, from the works of agricultural biotechnology critics one can deduce the simultaneous urgent moral imperative to help dependent foreign trade partners' achieve greater economic diversity and self-reliance.[43] Moreover, the gain to local self-reliance will presumably be slight if local communities simply shift their dependence from foreign markets to products and services provided by large domestic agribusinesses.

POLITICAL DECENTRALIZATION AND FEDERATION

Are there any other ways in which technologies can influence democratic self-governance? Positing a catch-all design criterion, grounded in Chapter 3's concern with decentralized political federation, may help identify some:

> Seek technologies that are compatible with globally aware, egalitarian political decentralization and federation.

This is Figure 5-1's Design Criterion G. To the extent that decentrally federated democratic politics may require, or be facilitated by, some technologies focally designed for that purpose, this is the first criterion that could strongly involve technologies' focal objectives.

Teledemocracy

The notion most commonly considered is that modern telecommunications and information technologies can profoundly affect political processes.[44] Many writers have noted ways in which electronically mediated political processes (such as instant voting or opinion polling) could, in the name of democracy, pervert political discourse or result in latent domination or mobocracy.[45] Others, often concurring, nonetheless advance plausible recommendations concerning alternative electronic technologies that could conceivably complement or strengthen decentrally federated democratic politics.[46] Important design subcriteria might include assuring universal, affordable access; individual privacy; a plurality of democratically governed communication modes and channels; equal and extensive access to government decision-making arenas; full representation for disempowered groups and communities within project design, implementation, and evaluation; and projects that encourage rather than enervate face-to-face egalitarian debate and decision making.[47]

For example, when a democratic local community already exists, there are intriguing possibilities for electronically sharing or merging its political discussions with elected representative or with the discussions of other communities. Thus town meetings could begin to function at regional, national, or transnational levels.[48] The possibilities are even more interesting if electronic media become democratically controlled (e.g., perhaps including competing teams of citizens from different cultural backgrounds filming and disseminating contrasting interpretations of the same events).[49] The computerization of corporate and bureaucratic operations also make it more technically feasible for outside groups to begin monitoring and participating in intraorganizational decisions that have public, structural implications.[50]

However, some circumspection is in order. First, one must not forget that technologies' democratic potentials rarely materialize without a fight.[51] Second, there are various ways in which seemingly decentralized electronic media could nonetheless facilitate centralized control.[52] Third, teledemocracy, even if ideally implemented, is no substitute for ensuring that the entire technological order—including the vast majority of technologies that focally have nothing to do with politics—becomes structurally compatible with democracy.

Finally, it is crucial to attend to the nonfocal dimensions of any technologies focally intended to support democratic politics. Telecommunications and information technologies are eminently capable of contributing nonfocally to democratically pathological psychological results.[53] Ideally one would proceed cautiously, working often on a small-scale, experimental basis, while paying careful attention to both focal and nonfocal consequences.[54] That is not how the construction of new telecommunication systems is currently proceeding.

Democratic Public Spaces

> It is the genius of the center city that it is *not* high-tech.
> —*William H. Whyte*[55]

Although teledemocracy is often accorded special attention, the foundation of democratic polities should include face-to-face politics in open, egalitarian social settings. For that, the most important technologies are not electronic communications technologies of any sort but the re-creation of more conventional public spaces[56]:

> I recently spoke to an audience . . . on the subject of the informal public life. I asked the group if Americans living in the suburbs had the freedom to put on their sweaters in the early evening and visit their friends at the neighborhood tavern. A resounding yes was given by the group. I asked if the younger children could go with coins in hand to the corner store and pick out gum or candy or a comic book. Another resounding yes. Finally, I asked if the older children could stop in at the malt shop after school. Yes was the response of the audience. . . . I'd hoped someone would realize that none of these people can go to a place that isn't there.
>
> If developers intentionally built communities without local gathering places and good sidewalks leading to them from every home, and did so for the purpose of inhibiting the political processes of the society, we would call it treason. Is the result any less negative without the intent?[57]

Prevalent enemies of democratic spaces include zoning laws that segregate homes from public gathering places; laissez-faire local development policies; the ethos of fast-food-franchise design, which intentionally discourages lingering; regional malls dominated by huge corporate outlets that sap city centers and police-out social spontaneity; and commercially abetted withdrawal into the privatized electronic cottages many call home.[58] In contrast, democratic places are where citizens gather naturally and informally for fellowship, solace, recuperation, release, diversity, surprise, ritual, performance, and politics. Examples include parks and playgrounds, inviting civic plazas, local pubs, skating rinks, shells and amphitheaters, street-level shops and cafés, convivial gaming rooms, museums and libraries, theaters, fountains and accessible waterfronts, meeting and assembly spaces, tree-lined promenades and arcades, neighborhood centers, and downtown marketplaces restricted to selling local or regional products in locally owned shops. Such places should be protected from the destructive effects of rampant commercialism and private autos; should be easily accessible by foot, bike, wheelchair, and public transit; and should be equally available to, and respectful of, all peoples' and classes' preferred cultural styles of interaction. To Lewis Mumford such places are "civic nuclei," to Benjamin Barber "talk shops," to Ray Oldenburg "great good places."[59] They are the heart and home of strong democratic politics.

Upon this foundation one can envision striving to engender complementary electronic spaces for geographically unrestricted forms of democratic deliberation.[60] Barber, for instance, warns that "the electronic town meeting sacrifices intimacy, diminishes the sense of face-to-face confrontation, and increases the dangers of elite manipulation."[61] Yet he finds sufficient offsetting benefits to merit exploring democratic variants of electronic public spaces.

James Fishkin proposes "deliberative opinion polls" as a hybrid method of retaining the virtues of face-to-face political deliberation within the context of a large society. He envisions nationally televising a representative sample of 600 U.S. citizens while they meet face to face for several days with competing presidential candidates. The citizens' discussions and final informed vote would, in turn, be intended to inform the wider, viewing electorate.[62] One can imagine more strongly democratic variants of this procedure in which the selected citizens would have more power to set their own deliberative agenda, would debate issues rather than candidates, and in which the process would be used explicitly to stimulate ancillary local deliberations nationwide.

Chapter 8

LIFE, LIBERTY, AND THE
PURSUIT OF SUSTAINABILITY

ECOLOGICAL SUSTAINABILITY

Strong democracy is grounded partly in an ethical duty to respect human moral agency by perpetuating democracy. It is well known, however, that technologies are capable of transforming ecological relationships in ways harmful to human well-being and survival. Insofar as biological survival is necessary to moral autonomy, and health is certainly helpful, it follows that it is democratically vital to:

> Avoid technologies that are ecologically destructive of human health, survival, and the perpetuation of democratic institutions. Seek instead an ecologically sustainable technological order.[1]

This is Figure 5-1's Design Criterion H.

Sustainability and Previous Criteria

Maintaining sensitivity to sustainability and its social preconditions is already implicit in the discussion in earlier chapters. Chapter 7 recommends that communities strive to ensure that their adverse ecological impacts remain locally concentrated. This prescription, coupled with relative local economic self-reliance, implies that citizens would experience more directly the ecological consequences of their technological activities and dependencies. Democratic empowerment should enhance groups' and communities' political capacities to avert ecologically destructive technological practices and systems. Greater local self-reliance reduces the energy consumption and pollution associated with long-range transportation. More self-reliant agriculture entails greater crop diversity; this diversity contributes to sustainability by increasing farm systems' innate resistance to pests and disease, thereby reducing their need to use nonrenewable resources and toxic chemicals to grow their crops. Finally, communitarian/cooperative technologies and con-vivial public spaces might enhance sustainability by substituting more satisfying and efficient collective consumption practices and common goods for the more profligate ways associated with hyperindividualism.[2]

Although quite helpful, these earlier criteria are not fully adequate to secure ecological sustainability. For that, Criterion H is necessary.

Paternalism

Chapter 7 also argues that while polities are morally obligated not to export significant translocal harms, individuals and local communities are not obligated to refrain from running informed risks to their own health or survival. Is this consistent with a duty to maintain ecological sustainability?

Yes, if one distinguishes again between harm to oneself and harm imposed on others. It is, for instance, democratically acceptable to go hang-gliding off a 3,000-foot cliff (assuming that you had opportunity beforehand to learn about the known risks involved), but not to engender ecological transformations that persistently or irreversibly damage a locality's hospitableness to human survival, health, and de-mocracy. Thus a democratic requirement in favor of sustainability qualifies, but does not override, Chapter 7's arguments against unjusti-fied political paternalism.

The Democratic Minimum

Is it therefore democratically impermissible to transform one's local environment? No. In fact, it is a conceit to imagine that one can avoid such transformation. "Nature" is itself a cultural category,[3] and people and their "natural" environments have always coevolved. It is accept-able to deplete nonrenewable resources and to damage common prop-erty resources (such as local fresh water supplies), but only to the extent that this does not harm other communities (Criterion E), one's own community's long-term health and survival as a democracy (including its capacity for relative self-reliance), and the local and translocal environment's hospitableness to future democratic communities and cultures.

Thus, so far as the minimum requirements for strong democracy are concerned, depleting a local nonrenewable resource is permissible so long as that resource will not foreseeably become democratically essential or provided that democratically satisfactory substitutes are (or will foreseeably and as accessibly become) locally available. Similarly, other forms of local ecological transformation—and even degradation, such as locally harmful pollution—seem democratically acceptable provided that those immediately affected concur, democracy's other necessary conditions remain satisfied, and firm provisions are made in advance to ensure that any degradation will in time be reversed at the expense of those causing or benefiting from the short-run degradation. To the extent that there is uncertainty regarding which resources might become democratically essential in the future, or concerning the reversibility of ecological degradation (taking into account future political readiness to actually implement a reversal), then short-run depletion or degradation ought to be restricted commensurately.

Finally, for the same reason that other communities or federation-level political institutions have an obligation to oversee local compliance with the categorical imperative in the present, they may also oversee local compliance with communities' collective responsibility to future generations.[4] Likewise, the collectivity of local communities and voluntary associations ought to strive to ensure that federation-level policies and activities support long-run sustainability. On the other hand, in the context of today's worldwide structural inequalities, affluent communities or nations would seem to have little moral right to insist that poorer counterparts assume the additional costs or austerity measures associated with sustainability.[5] Instead, one might argue that affluent societies are morally obligated to try both to redress intersocietal structural inequities and to assist other societies in achieving sustainability.[6]

Beyond the Minimum?

What does it mean that the preceding requirements represent the democratic "minimum"? The task of democratic theory is to strive to specify a set of minimum conditions needed to establish and perpetuate democracy. Beyond that, it is up to citizens, cultures, and societies to decide for themselves questions such as whether their moral obligations to other species or the world extend past that minimum level.[7]

While not requiring more than the minimum, strong democratic theory of course permits deeper ecological sensitivity. Democratic egalitarianism would also provide conditions under which those favoring heightened sensitivity would have fair wherewithal to press their case.[8] Moreover, democracies ought to invoke especially cautious and respectful procedures when places, objects, or values sacred to some people are jeopardized.[9] This procedural caution would provide further assurance that the concerns of any person or culture that finds an environment sacred would receive a full and respectful hearing.

One might even speculate that as more people lived in closer contact with natural rhythms (as would tend to happen as a by-product of greater local economic self-reliance)—and yet retained the inescapable modern insight that ecological devastation is within our technological capability—the proportion of people who would experience a more reverential regard for the transhuman world would increase.[10]

Concern with individual psychological development and self-actualization could conceivably provide additional democratic grounds for sustainability, and specifically for preserving wilderness. It might prove possible, for instance, to derive from the philosophy of Erazim Kohák a democratic duty to prevent the world, or even local environments, from falling fully under the sway of human design impulses.[11] That is, suppose that a fully designed world, as opposed to one merely influenced by people, were so culturally impoverishing as to radically impair individual moral growth and self-realization? Analogous reasoning, elaborated this time from Albert Borgmann's thought, might suggest a democratic duty to lower reliance on the kinds of technologies he calls "devices" on the grounds that, by divorcing people from experiential engagement with the natural world, they impair the motivation needed for a deep and enduring commitment to sustainability.[12] Both of these elaborations invoke psychological hypotheses that merit deeper investigation.

Compared with this volume's previous design criteria, Criterion H—while more stringently formulated than most contemporary environmental policies—is nevertheless the one criterion that has achieved fairly extensive legitimacy in the modern world. It is also the one criterion that is the least specifically democratic. That is, democrats must be concerned to perpetuate the ecological conditions of human survival and health, but dictators who want their memory to endure might well manifest a similar concern. For instance, Theodore Roosevelt–era U.S. conservationists were often profoundly antidemocratic, and modern Singapore has adopted some exacting environmental policies although its government is authoritarian.[13] In other words, if sustainability is abstracted from a broader democratic vision, it is consistent with reproducing virtually any system of social relations, including democracy, but not excluding others. Thus the one, not purely economic design consideration that contemporary politics most readily includes is still not explicitly democratic.

In short, ecological sustainability is a necessary but insufficient material condition of strong democracy. In turn, strong democracy morally entails and promises to facilitate achieving sustainability.

STRUCTURAL FLEXIBILITY

There are many ways a society can be organized as a strong democracy; no one set of institutions or technologies is best for all times and all places. Consequently, societies ought to preserve a capability to modify their structures in case they prove insufficiently democratic, to adapt to new circumstances, or simply because they want to. Conversely, they ought to be able to resist forces that would compel them to undergo structural transformation that is substantively undemocratic or that they otherwise oppose. In the United States such needs are reflected in the Constitution's provision for amendment and in the procedures governing the evolution of legal statutes and precedents.

Technologies bear on these issues both because they are social structures and because they influence other social structures (just as those others influence technologies). One might, moreover, expect technologies to prove especially significant with respect to structural flexibility, because they are partially embodied in recalcitrant physical materials and sometimes organized into large systems.

In short, it is democratically vital to design technologies and technological orders for compatibility with ongoing democratic structuration.[14] This ought to encompass design-compatibility with democratic procedures able to transform technologies (and other structures) in substantively democratic ways, while averting antidemocratic structural transformation. This objective is elaborated into Figure 5-1's Criterion I in two stages.

Local Technological Flexibility

Technologies exhibit a margin of flexibility in their range of focal and nonfocal results. However, some technologies are more flexible than others, and some in democratically more significant ways. What happens if a technology is adopted through a democratic process and is substantively democratic, but citizens decide it no longer suits their needs? Suppose, contrary to prior expectation, a deployed technology now appears undemocratic?

Within the context of an otherwise democratic sociotechnological order, a few deviant technologies may not matter. Otherwise, perhaps the offending technologies can be abandoned or replaced. But if one is dealing with a large, tightly integrated technological system, or if a technology has grown to occupy an economically or culturally central position with other social structures now closely adapted to it, then structurally consequential modification could prove difficult. Such possibilities suggest that as a technological order is democratized, it is important to ensure that it develops in ways allowing for future democratic flexibility.

Limits

Are there moral or practical limits on how far a society should go toward achieving structural, including technological, flexibility? Whereas in principle a people could change their minds frequently, it seems unrealistic to imagine that every month, or even every decade, an entire technological order could be reconstituted comprehensively. Consider, too, the perspective of children. They are born into a world in which myriad technological commitments have already been made. How can they ever experience the inherited technological order as an expression of their own free will? That order cannot conceivably be revolutionized structurally each time a group of children attains the age of reason. As Marx noted, "Men make their own history, but they do not make it just as they please . . , but under circumstances directly found, given and transmitted from the past."[15]

What seems required is a judicious balance between, on the one hand, chaotic or rampant structural change (the existential equivalent of no structure) and rigid (especially rigidly antidemocratic) structures on the other. The extreme of "no structure" fails to provide opportunities to develop autonomy or fulfill life plans.[16] The alternative extreme of authoritarian structures does the same. Thus it seems we require democratic structures that permit democratic modification, while at the same time providing some resistance against antidemocratic transformation.

Creating a relatively mutable technological order does not mean societies should necessarily exploit all that potential mutability. Maintaining the potential to alter shared life circumstances—or to allow them to evolve spontaneously, subject to democratic oversight—would help people feel emancipated through their technological involvements. Yet a technological order that is, in practice, relatively stable can provide contextual continuity constitutive of coherence, meaning, a measure of predictability in human affairs, and moral growth. Without some degree of technological stability individuals might, upon approaching the "no structure" extreme, become disoriented or incapable of sustaining stable identities. Far from becoming autonomous selves, they might thus fail to become "selves" at all.[17]

Excess flexibility might also produce an ephemeral, culturally impoverished world. Ancient pyramids, Gothic cathedrals, the Taj Mahal, and many other grand historic structures were not produced democratically. But in principle they could be, and in other respects the results remain magnificent, spiritually uplifting, and vital to the lives of their surrounding cultures and communities. Could there be no legitimate place for works of such enduring grandeur and emotional force in democratic societies?

Perhaps one solution is to recognize that most such structures are compatible with multiple, shifting social uses and relationships, and thus are not—social structurally speaking—so inflexible as one might suppose. For instance, 12th-century French cathedrals functioned initially as churches but later also as schools and as meeting places for guilds and civic assemblies.[18] Moreover, unlike many modern, large integrated technological systems, large architectural edifices and civil engineering works can, if compelling reason arises, generally be demolished without disrupting the technical operation of other, far-flung sociotechnological systems. Bijker and Aibar even describe the case of an enormous and complex, modern Dutch dam that underwent successive significant episodes of politically guided redesign during its prolonged construction.[19] Perhaps, finally, flexibility is particularly important when dealing with radically novel technological innovations.[20] That is, after an innovation has been in use a long time in various contexts, and provided there has been careful democratic scrutiny of its current and potential social structural effects, less flexible variants might become permissible.

There is reason to hope that, under the influence of a reasonable compromise between structural rigidity and infinite flexibility, children, as they mature, would become progressively more able to recognize their sociotechnological order as a ground of social freedom, gradually assuming their role as conscious agents of ongoing democratic structural reproduction or transformation. Meanwhile, adults, beyond increasing the ability to develop and express their own autonomy, would be able to discover the moral satisfaction of knowing that they had bequeathed succeeding generations conditions under which their freedom, too, could flourish.

This should satisfy the psychological concern that individuals need a symbolic sense of immortality to find satisfaction and develop autonomy in their lives today.[21] Within a democratic sociotechnological order, symbolic immortality would not be embodied in rigid social structures that might subsequently weigh "like a nightmare on the brain of the living,"[22] but in bequeathing an array of flexible, autonomy-supportive structures that descendants could, with time, appropriate and make over as their own. In other words, a sense of immortality would emerge from the moral practice of respecting future generations' capacities for autonomy, just as they could come to respect today's societies for having done so and, furthermore, continue the legacy by passing on another set of democratic structures to generations to come. Kant's categorical imperative would, in one sense, extend intertemporally in both directions.[23]

This establishes the first half of Figure 5-1's Design Criterion I. Societies should:

> Seek technologies and a technological order that harbor extensive, potential structural flexibility (remaining, however, as much as possible within the broad domain of potentials that are substantively democratic).

Dimensions of Flexibility

Flexibility can be defined along many dimensions. However, one concern must be to escape bedazzlement by displays of technological change or flexibility that latently reinforce existing structural relations.[24] For instance, many seemingly radical technical innovations in manufacturing conform closely to preexisting patterns of hierarchic work relations.[25] Similarly, many new electrical household appliances, setting aside their social effects within the home, have done nothing to transform the established pattern of relations between households and the electric utility system. Thus we will concentrate here on structurally consequential, often nonfocal, dimensions of flexibility. A first-cut prescriptive list might suggest that, all else being equal, societies seek technologies that accomplish the following:

1. Require active maintenance of only relatively nonspecific, geographically nonextensive background conditions. Otherwise, maintaining background conditions can evolve into an imperative that reduces social flexibility. This is problematic, especially if many people or communities depend on a technology or if some of its background conditions are largely beyond the influence of individuals or communities (tending to violate also Criterion F's counsel regarding self-reliance). In this sense, using a hammer makes fewer and less specific demands than using a refrigerator, telephone, or automobile.[26]

2. Require few and relatively nonspecific inputs (eschewing locally scarce, nonrenewable resources or use of renewable resources in excess of locally sustainable yield).[27] Likewise, a technology that stresses ecosystems near to their sustainable carrying capacity reduces social maneuverability.

3. Permit a wide range of self-actualizing activities and nonauthoritarian social relationships. In this respect large, tightly integrated technological systems that could engender social calamity if they malfunction (e.g., air traffic control systems or nuclear power plants) are particularly problematic. Such features may make it seem too risky to permit the kind of organizational experimentation needed to discover whether a system harbors latent structural flexibility.[28]

4. Permit producing a relatively wide range of substantively democratic material outputs, services, or amenities.[29] Programmable computers and electronic communication media exemplify extraordinary flexibility in this regard. (However, flexibility in focal performance does not necessarily correlate with flexibility in other structurally significant dimensions. For instance, there are variants of advanced telecommunications networks that would be consistent with highly undemocratic systems of social control.[30])

5. Can potentially operate in conjunction with a wide array of parallel or complementary technological practices, other social activities, and other democratic social structures.[31]

6. Can be altered with respect to criteria 1–5 relatively quickly and inexpensively.

7. Can be planned and built relatively quickly—or modified creatively in the course of planning or construction—and could, if desired, be abandoned or replaced quickly and inexpensively. Very large-sized technological units (such as nuclear reactors or aircraft carriers) and large, tightly integrated technological systems (such as some systems of urban infrastructure) warrant particular scrutiny. The long lead time to plan and construct such technologies can favor centralized technocratic planning over democratic participation.[32] Long lead times can be unforgiving if new social exigencies arise after construction is underway.[33] Long lead times can also allow planning forecasts to congeal into self-fulfilling prophecies that foreclose future options. For instance, the U.S. interstate highway system, planned when only about half of all American families owned a car, helped kill off competing mass transit systems; this virtually compelled later, near-universal automobile ownership.[34] On the other hand, medieval Gothic cathedrals often took a generation or more to construct yet permitted significant modification and creative expression along the way.[35]

The idea that technologies ought to be relatively easy to abandon or replace is not a prescription for shoddy construction, disposability, or planned obsolescence. The latter practices represent pseudoflexibility because they preclude the option of not abandoning. They also squander natural resources and generate waste, thus tending to violate earlier design criteria regarding ecological sustainability and externalities.

8. Create relatively few and insignificant consequences that are potentially adverse (especially democratically so) and that could be irreversible, persistent, or very costly to reverse or mitigate.[36]

This suggests, all else being equal, that a furnace that can burn any combination of coal, oil, gas, wood, or garbage is better than one that only burns one fuel. The Volvo auto factory at Kalmar that permitted frequent restructuring of assembly operations was preferable to a standard auto assembly line. An appliance that people can fix themselves or take to a local shop for repair is better than one they have to discard or return for repair to a distant factory. Pollution that biodegrades quickly is preferable to chemical contamination that will persist for generations.

Perhaps further inquiry will frame the concept of flexibility more precisely. But extensive democratic interpretation will probably remain required to determine, especially when there would be a price in terms of other democratic opportunities foregone, how much flexibility of which kind ought to be structurally embodied within a technological order.

Global Technological Pluralism

Beyond flexibility, a second defense against becoming imprisoned by past technological choices involves maintaining a societywide (or "global") pattern of cultural and technological pluralism.

Chapter 3 provides several arguments in favor of cultural pluralism. Technologies, on the other hand, play a constitutive role within culture.[37] It follows that it is democratically vital to ensure that cultures are able to maintain a relatively autonomous capacity to develop and perpetuate their own style of technological design and practice.[38]

Requisite conditions for maintaining cultural pluralism include striving through translocal cooperation or decentrally federated politics to ensure that no morally tolerable culture is deprived of fair access to the resources needed to perpetuate itself as a living tradition; that no cultures can impose themselves hegemonically on others through force, ideology, or other democratically illegitimate means; and that translocal social structures and dynamics remain subject to federated democratic oversight and, as necessary, control.[39]

With respect to technology, this implies each culture's right to develop and preserve a distinctive technological style. Through federated democratic discourse, ideally reflecting the influence of intercultural citizen sabbaticals, cultures can assist one another in recognizing technologies' nonfocal structural significance and learn to adopt a critical attitude toward imported technological styles. The purpose is not to block cultural evolution or technological transfer, but to ensure that cultures develop along paths of their own choosing, preserving from the past that which they consider essential. Through translocal or intercultural federated politics, local communities and cultures must also be empowered to resist technological imperialism, be it a direct expression of the disproportionate power of other cultures, polities, or translocal organizations, or via other relatively uncontrolled sociotechnological processes.[40]

Similarly, as new cultures emerge or as oppressed cultural groups gain relative autonomy, they must each be allowed to evolve their own technological voice—a technological style consonant with other elements of their culture. This might entail preventing aggressive entrepreneurial intrusion, facilitating intercultural redistribution of resources, and establishing mutualistic intercultural sharing of know-how in ways that respect the moral equality, knowledge, and creativity of each participating group.[41] Under the present, highly inegalitarian global political and economic circumstances, the need to find practical means for assisting disadvantaged societies' and communities' struggles for cultural integrity is urgent.

The force of these arguments should not be belied by their simplicity. Acting on them might entail reconceiving modern Western technology and science-based culture as just another kid on the block: a set of traditions and practices not superior to others from the standpoint of any universal criteria—perhaps even inferior in terms of their failure to comprehend technologies' polypotency and structural significance—and as likely as any other cultural practices to benefit from mutually respectful cultural interchange.[42]

Pluralism would guarantee that if a community's sociotechnological order is not to some citizens' liking, but they are unable to persuade fellow citizens to change it, they would always be able to search for another community that more adequately realized their ideals. Permanent migration ought normally to be an option of last resort. However, together with flexibility within each community, pluralism among a society's (or the world's) many different communities would constitute a robust system of safeguards against radical mismatch between present desires and past technological choices. Hence Figure 5-1's Design Criterion I:

> Foster local technological flexibility within a global pattern of democratic technological pluralism.

The democratic benefits of cultural and technological pluralism are, however, broader than just indicated. Each democratic culture benefits from the coexistence of many others by being able to debate a more diverse range of alternative technological designs, through enhanced ability to unmask ideologies propagated about or through technologies, and in opportunities for self-realization. (For instance, Chapter 4's discussion of the Mundurucú helped dramatize the authoritarian and individualized nature of modern technological orders.) Criterion I also implies that processes for popular participation in technological decision making, as prescribed in Chapter 3, ought to respect the rights of citizens from different cultures to speak and act autonomously rather than in terms of any hegemonically imposed language, agenda, or institutional framework.

A residual question is whether democratically prescribed forms of intercultural interaction (e.g., citizen sabbaticals, transcommunity technologies, democratic federation, or permissible intercultural trade) would tend to subvert and homogenize cultures. Probably not, in part because the question tends to assume falsely that non-Western cultures were traditionally static and insular. Viable culture is a living process and an evolving set of structures, continuously rewoven and re-created through the actions of its members. Most of the supposedly insular cultures that Western social scientists studied in the past probably not only engaged in lively intercultural interaction, but had specifically (although sometimes to their regret) interacted with European culture long before trained anthropologists arrived on the scene.[43]

Cross-cultural citizen sabbaticals promise to enhance mutual comprehension and empathy, thereby contributing to intercultural tolerance and support. One necessary proviso, however, might be that communities and cultures must have some right to protect themselves from inundation, perhaps by setting a limit on the number of temporary residents permitted at any one time. For instance, the Hopi Indians living in traditional mesa villages in the U.S. Southwest have learned, in the interest of preserving their way of life and sanity, to impose careful restrictions on tourism.

In general, cultures and local communities have much to gain from membership in a broader democratic society or from situation within a more strongly democratic world order. Assuredly, current global economic, political, and technological forces are an ongoing threat of major, sometimes calamitous, proportions. However, the threat of cultural uniformity or destruction comes not from intercultural interaction per se but from interaction on unequal, nondemocratic terms.

Toward a Democratic Politics of Technology

Part II developed a provisional system of contestable design criteria for identifying democratic technologies. Part III explores the idea of a democratic politics of technology. Are we to continue as the Ibie-canesque victims of arbitrary or antidemocratic technological forces, or can democratic design criteria help us evolve capabilities closer to those of the Amish in comprehending and guiding our own fate?

Chapter 9

"THINGS LIKE THAT IS PRICELESS"
Contestability and Dignity

USING THE CRITERIA

The criteria developed in Part II can help people perceive technologies more holistically and become more aware of a technological order's structural significance. In that way democracy can more easily find its preeminent place among the issues that today dominate technological decision making.

Technologies and Technological Orders

If one reexamines the design criteria listed in Figure 5-1, it becomes evident that all of them apply to particular technologies, to their evolving array of supporting background conditions, and to technological orders as a whole. For instance, Criterion A suggests that citizens judge whether a particular technology involves authoritarian versus democratic social relations. However, the criterion also requires maintaining a balance within each local technological order among communitarian/cooperative, individualized, mass, and transcommunity technologies. Maintaining this balance entails evaluating many different technologies, considered as an integrated whole. Chapter 4 noted, for example, that contemporary norms of individualism and privacy are reproduced through complexes of artifacts and practices that structurally complement one another.

A System of Criteria

The aforementioned balancing involves using one criterion at a time. However, technological evaluation needs to incorporate all of democracy's necessary conditions. For instance, sometimes a technology that is satisfactory in terms of several criteria can be harmful in terms of others: Volvo's most ambitious attempt to develop a convivial, skill-enhancing automobile production plant was initially opposed by environmentalists on ecological grounds.[1] In such cases, citizens will have to decide whether the democratic pros outweigh the cons, or whether there are alternative designs that would improve a technology's overall consistency with democracy.

Criteria in Context

Furthermore, technological judgment must take into account a range of contextual considerations. One can't generally make a blanket claim such as: "Windmills are democratic." The political significance of windmills varies according to their types and their settings. Consider a society where electricity is provided by a few large, fossil-fuel power plants operated by a hierarchically managed, regional utility company. There a local community might well enhance its material possibility of democracy by deploying a system of community-owned and -managed electricity-generating windmills.[2] On the other hand, consider agrarian communities in which common interests are reproduced in part through collective management of a community-scale dam and canal irrigation system.[3] In that context, replacing the system with individually owned and controlled, windmill-pumped water wells could undermine an important basis of community solidarity, damaging the sense of commonality useful for democratic self-governance. Thus, generally one can only assess whether a technology is democratic within a specific context.

Moreover, Part II articulates criteria at a high level of generality, precisely so that they apply in disparate circumstances. But that means that using them contextually will typically demand further elaboration and judgment. This is evident, for example, in Part II's discussions of virtual community, automation, and teledemocracy.

A Fluid Medium

The concept of contestability suggests that democratic design criteria can only function constructively if they remain socially interpreted as a fluid and flexible political language. If they degenerate into rigid, mechanically interpreted categories, the results will be procedurally and substantively undemocratic.[4] Could anything be further from strong democracy than fat bureaucratic tomes purporting to demonstrate "Formal Compliance with Subcriterion 28.1(e) of Design Criterion 574" or than committees of experts informing citizens of things that, in reality, only citizens can discover and decide for themselves?[5] Societies and technological orders are plastic entities. An overly rigid system of design criteria would preclude the subtle adaptations and artful judgments that reproducing democracy and expressing freedom both require.

One means to preserve openness and flexibility is to ensure that democratic contestation can incorporate more than formal argument, by encompassing interpretation and evocation as well.[6] This implies honoring a broad range of communicative means and media, including facial expression, body language, stories, poetry, dance, myth, drama, art, and technological performance itself.[7] Like the parables of ancient Chinese philosophy, evocation does not strive via logic to compel direct assent, but instead establishes an experiential context in which participants can discover their own individual and shared insights.

Incompleteness

The criteria proposed in Part II are "general" in the dual sense of aspiring to apply to all societies and to all technologies.[8] There is no reason to think these criteria are, as a system, complete.[9] For example, securing a democratic technological order will require additional general criteria (e.g., concerning civil liberties), less general criteria focused on particular focal classes of technologies (e.g., biotechnology and military weaponry), and other criteria developed locally for specific application to democratically salient local circumstances.[10]

DEMOCRATIC POLITICS AND THE SACRED

Chapter 3 argues that strong democracy is a "highest order shared human value." As such, it should arguably take precedence over other considerations in political deliberations. But there is ambiguity here: are there other highest order values that are not shared?

Of course. For some people these can be values associated with organized religion, a spiritual or mythic tradition, or a personal god, goddess, or gods; for others, a sense of nature or particular places as sacred; for still others, an overriding commitment to family, relationships, a vocation, or some other personal or social ideal. Experienced as supremely or eternally valuable, these are things that confer special dignity, meaning, and purpose on our lives.

Reasoned argument suggests that strong democracy is a "highest order shared" value insofar as it is necessary to fully realizing individual moral autonomy, dignity, self-worth, and the enjoyment of other goods. In contrast, other supreme or sacred values depend on personal experience or religious faith. They are supremely significant to those who uphold them but are not authoritative for—and cannot rightfully be imposed on—those who do not.[11] Indeed, were it otherwise, whatever one might personally experience as sacred would never be secure against contrary political dictates. (This is a central reason that the U.S. Constitution requires separation of church and state—not because the state is superior to religion, but in order to protect freedom of religion from political interference.)

Thus, if strong democracy and moral freedom are distinctive among supreme values in the extent to which they are shared,[12] that is in part because democratic morality requires establishing social or political means for respecting other supreme values that may not be widely shared. To do less risks disrespecting and diminishing the lives of those for whom these values are supreme, which would violate strong democracy's central norm of unqualified respect for the moral autonomy and dignity of all people.

Technologies enter this arena because they can symbolically embody sacred values or threaten sacred objects, settings, and practices. For instance, virtually all cultures locate sacred worth in various artifacts, including objects used in religious rituals, certain works of art and architecture, or material symbols of social unity such as flags. Conversely, technologies are eminently capable of jeopardizing sacred values, places, things, or relations, and this capability plays an unmistakable, if often latent, role in deciding which technologies become embroiled in public controversy. This appears true of disputes such as those concerning the fluoridation of public water supplies, human reproductive technologies, genetic engineering, artificial intelligence, toxic waste disposal, or the siting of many large industrial facilities.[13] For instance, polemical literature concerning nuclear energy, authored by scientist and layperson alike, is rife with religious and mythical imagery: Promethean fires, Pandora's boxes, Faustian bargains, manifestations of Krishna, genies unbottled, and fruits forbidden.[14]

Whenever sacred values are at stake, democratic processes ought to redouble the normal effort to accord equal and respectful weight to each person's or culture's concerns—at least unless it is democratically established that one or another group's practices significantly jeopardize democracy in the broader society.[15] Furthermore, respecting each person and culture entails affirming their right to disapprove of formal analysis or quantitative assessment of the costs and benefits of a proposed action.[16] The sacred is a domain of transcendent or eternal value; sacred things are literally priceless:

> Everything has either a *price* or *dignity*. Whatever has a price can be replaced by something else as its equivalent; on the other hand, whatever is above all price and therefore admits of no equivalent has a dignity.[17]

In this instance, Immanuel Kant captured a high philosophical idea that is not alien from worldly experience. Two centuries later one can find a man from an impoverished Appalachian coal-mining community expressing virtually the identical sentiment. Here he explains the meaning of the personal objects that he lost when a dam burst and washed away his home:

> Well, I'll tell you. I had a lot of stuff I wouldn't have traded the world for, and I just didn't declare them [for an insurance claim] at all. I had an old shotgun my daddy gave me when I was just a kid. It wasn't worth more than ten or twelve dollars when it was new, so what's it worth now? I had a silver dollar minted the year my daddy was born, 1888. He gave that to me, too, and his daddy gave it to him. What's it worth? A dollar? Things like that is priceless. You just can't ever get them back.[18]

Imputing a price to something sacred can jeopardize or annihilate its supreme worth. The same holds true if one analytically sunders something sacred into lesser parts.[19] And while it is always democratically vital to permit people to frame issues in a style and language of their choice, this is especially true when the sacred sphere is involved; people must be able to comport themselves in ways that demonstrate appropriate respect for objects or meanings they hold sacred. All of these stipulations are commonly violated in contemporary technology assessment processes and in judicial and administrative hearings concerning technologies.[20]

Such procedural guidelines can affirm the moral worth and dignity of each person or culture. They are also more likely to elicit participants' reciprocal respect for the procedures and for other participants.

Moreover, while people do not normally analyze things they hold sacred, nor exchange them for mundane things, they do sometimes trade one sacred thing for another. A man who would never sell his grandfather's gold watch to buy a television set would probably sell it readily to help pay for a life-saving operation for one of his children.[21] Thus, to the extent that democratic procedures and their participants elicit sufficient respect to earn a kind of sacred status of their own, participants may become willing to entertain compromises—trading off their initial sacred valuations against newly acquired ones—that would previously have been unconscionable. The resulting increased level of mutual respect and flexibility would enhance the prospects for achieving consensual democratic decisions. However, even if consensus remains elusive, only when all parties feel respected by a process are they in turn warranted in honoring it and its outcomes—even particular outcomes they disfavor.

Chapter 10

CHEESEBURGERS, DEODORANT, AND THE BASIC STRUCTURE OF SOCIETY

Democracy versus Economics-as-Usual

Democratic politics of technology may look promising on paper, but in the real world of technology economic reasoning prevails. Can one seriously expect that to change? Yes.

Whenever technologies structurally affect democracy or other important shared values, democratic politics ought to take precedence over economic calculation or unregulated market outcomes. This implies that contesting democratic design criteria can be superior to using today's formal methodologies for public policy analysis. Those methodologies—known by such names as "welfare economics," "cost–benefit analysis," and "risk assessment"—ultimately derive from utilitarian ethics and neoclassical economic theory.[1] Insofar as each is concerned with selecting efficient means to ends not seriously examined, they can be called "economistic."

Although often criticized,[2] these economistic methodologies provide today's dominant languages for considering public policy, especially policy regarding technology. Practitioners commonly retort: "Agreed. Our methods are imperfect. We are working to improve them. In the meantime, what would you prefer: dogmatism, gerrymandering, chaos? What's your alternative?"

This is a fair question. The alternative proposed here—a democratic politics of technology—pays close attention to the social struc-tures within which everyday life, including economic life, unfolds. Although concerned especially with technological structures, this ap-proach may have broader significance insofar as it provides a vantage point for an overall critique of economism, thus potentially initiating a comprehensive alternative.

OMISSIONS AND CONFUSION

Philosopher John Rawls refers to the background conditions of everyday life as the "basic structure" of society.[3] These conditions are significant because of their role in patterning social relations and in developing individuals' capacities, preferences, needs, and identities. However, whereas Rawls treats only laws and dominant governmental and eco-nomic institutions as constituents of the basic structure, here additional constituents include language, cultural belief systems, and technology.

The ways in which technologies influence preferences and needs, for example, is richly diverse: highways and dispersed housing patterns contribute to the widespread need to own a car; owning a car in turn establishes a need for gasoline and repair services. U.S. highways have also, by enabling businesses and more affluent families to move to the suburbs, contributed to the creation of today's desperately deprived urban underclass; fear of urban crime has in turn led affluent families to seek high-tech fortified houses and housing developments.[4] Business partners' or competitors' purchase of a new communications technol-ogy, such as fax machines, help make it necessary to buy one's own (or else be left out). Teenagers crave new clothes and accessories thanks partly to the influence of print and broadcast media; they also want to emulate their cohort's acquisitions. Ibiecans may decide that they want television sets when, thanks to pipes, washing machines, and air condi-tioners, they no longer share a vibrant community life.[5] When scientists implicated atmospheric emission of CFCs (chlorofluorocarbons) from air conditioning, styrofoam production, and other industrial applica-tions in depleting the earth's ozone layer and consequently increasing risks of skin cancer, this induced new consumer demand for wide-brimmed hats and sunblocking lotions. Generally, many vague yearn-ings acquire their specific content in terms of the range of products and services that are technically available.[6] These are ways in which tech-nologies, as one aspect of their structural role, affect our psychological makeup.

Economists, however, are often oblivious to social structures.[7] Economists' models conflate needs and preferences, treating the latter as "exogenous"—generated independently of economic interaction or institutions—and even static.[8] Economistic methodologies are concerned with satisfying individuals' preferences within a taken-for-granted social order. As an introductory textbook explains, "Policy analysis is . . . a discipline for working within a political and economic system, not for changing it."[9]

Some critics of economism have challenged the assumption of exogenous want formation on the grounds that an individual's own economic actions are psychologically transformative.[10] That is, as individuals act, they develop new skills, insights, interests, and preferences. For example, by working at a variety of jobs, people learn about the kind of work they like and dislike and often change psychologically in the process.

Other critics note that economic activities such as advertising and marketing help transform people's wants. For instance, Procter & Gamble developed and marketed Crisco vegetable shortening early in the 20th century not to meet preexisting consumer demand but to guarantee access to the large quantities of cottonseed oil that it needed in manufacturing other products, such as Ivory soap.[11]

> In creating the techniques to make people want things, marketers developed principles that belied neoclassical economic theory. According to that theory, price—determined in the marketplace by supply and demand—functions as an information feedback system, telling producers how much of their product to make. . . . In actual practice, manufacturers operated on the new principle that demand could be created by the manufacturer.[12]

These observations constitute additional departures from economism. However, as bases for criticizing the treatment of preferences as economically exogenous, they remain incomplete unless combined with the previously noted, largely unplanned, psychological influence of social structures.

In a further descriptive departure from neoclassical economics, the critique developed here begins to diverge from most critics as well, ranging from neo-Kantians (such as Rawls) to radical political economists. Perceptive neoclassical economists—those who acknowledge the role of a basic structure in maintaining economic institutions—generally regard it as a realm apart from daily life. By tacitly accepting the view that a society's structure is rebuilt only during times of war, revolution, or constitutional convention, critics unnecessarily marginalize themselves, for they can thus only hope to be influential during crises that provoke societal readiness to entertain heroic questions.[13]

A society's basic structure does not originate and reside merely in government organization but also in other systems of social structures and in the everyday activities and beliefs of a society's members. People are the evolving products of structures that they help continuously to reconstitute. This is structuration.[14] Within the present context, the principal structures of concern are technologies and technological orders. It is, in contrast, consistent with economism's general insensitivity to structures and structuration that it treats technologies—whether categorizing them as consumer goods, factors of production, or infrastructure—as though they did not produce structurally significant psychological and other effects.

Whoops: Structures as "Externalities"

One can interpret structuration using economism's own concept of externalities.* This reveals, in a way complementary to ideas expressed in Chapter 3, that there is little danger of democracy interfering with the efficiency of competitive markets because—quite independently of politics or of economic concentration among producers—market efficiency never obtains anywhere. This is true of efficiency in two senses: in its technical sense (of there being no way to make any person better off without making at least one worse off—so-called Pareto optimality), and in the looser sense of providing people, subject to resource constraints, with roughly the commodities and the world they want.[15] (There is a third sense in which markets are reasonably efficient: they tend to clear. Goods and people move about with fewer of the bottlenecks characteristic of command economies that suppress self-setting pricing mechanisms. Whether the results of market clearing are remotely consistent with peoples' wishes or with democracy's necessary conditions is quite another matter.[16])

Let's begin by adopting the ordinary economistic assumption that people's preferences are static. This assumption is false but generous; relaxing it would reinforce the argument against economism. Next observe that in real life there are no such things as *pure* private goods.[17] Every commodity or event displays *spillover* effects or externalities; some

---

*This section uses a bit of economic terminology, but it may be skimmed without losing the overall train of my argument. There is no agreed-upon, precise definition of externalities; for our purposes it suffices to think of them as the effects of an economic transaction on people not party to the transaction. When you walk down a street and smell pollution not coming from your own chimney, that is an externality; so is virtually everything else you experience during your stroll, including the clothing and behavior of passers-by. See also Chapter 7.

are anticipated and some are not, ranging from the aesthetic, psychological, and biophysical to the ecological, the economic, and the political.[18] Indeed, competitive markets often reward firms for displacing costs onto other people in the form of externalities. Externalities are not, as some economists imagine, a rarity; they are utterly pervasive. But economists who recognize this truth nonetheless fail to comprehend its full significance.

First, insofar as it is inconceivable that externalities, being pervasive, could ever all be evaluated and internalized via administered shadow prices, one may deduce that economic efficiency—in neoclassical economists' strict and preferred sense of Pareto optimality—can never obtain in the real world.[19] (An administered shadow price attempts to offset market failure by forcing producers to pay for the external harms they cause.)

Second, the so-called "general theory of second best" states that when—as I argue is ubiquitously the case—there are externalities or other market failures, then there are no defensible, a priori welfare–economic grounds for seeking to establish the remaining conditions required for perfect competition.[20] There may often be commonsense, empirically informed reasons to do so,[21] but these do not preempt deeper moral and political considerations discussed momentarily.

Third, externalities are frequently coercive or seductive.[22] Representing more than a static deviation from conditions of Pareto optimality, externalities alter the social context in which they appear, inducing people to experience and behave differently. Generally, this altered behavior includes producing new externalities, sometimes even more of the initial externality, establishing a positive feedback loop that can swiftly transform an entire community or society. Chapter 2 offered as an example the manner in which a neighbor's purchase of a noisy, power lawn mower can influence other neighbors to do the same. History reveals countless additional cases. For instance, as wealthier Americans acquired automobiles during and after World War I, some shops began eliminating delivery service and moving out of residential neighborhoods. This represented an external effect that encouraged or forced still more families to buy cars.[23] The cycle then continued, as remaining businesses were gradually compelled, in order to keep costs in line with those of their competitors, to eliminate their delivery services. Thus, in time, delivery service all but disappeared, even though no one involved—not even the first wealthy people who sometimes shopped by car—may consciously have wished for that aggregate result. In a similar vein, the currently anticipated expansion of home electronic shopping and service delivery can be expected to force some existing stores and service providers out of business—even if no consumers wish that to happen—thus inducing or forcing more people to shop electronically.[24]

Drawing on ideas developed by Borgmann,[25] one might further conclude that each decision to purchase a new device—be it a telephone, a compact disk player, or a kitchen appliance—tends to have the unintended spillover effect of contributing to evolving a world in which there are weakened family ties and less local community life or routine personal engagement in nature. Instead, people grow ever more dependent on—yet experientially disengaged from—vast, impersonal supporting background systems (such as regional electricity grids and the global petroleum industry). Eventually, ensuring that such systems are maintained can become a practical imperative for national governments, significantly affecting their agendas and institutional structure.

Adverse externalities can also perversely inhibit adoption via the market of alternative technologies that would reduce production of the initial harm. For instance, the pollution, noise, and danger of cars inhibit the alternative choice to commute by nonpolluting bicycles. The safety risks associated with nuclear power stations and the pollution caused by coal-burning electric plants likewise reduce individuals' incentives to conserve energy or to adopt solar technology. In other words, unless coordinated politically, such individual steps do nothing to lower the individual's own exposure to the externally generated risk or harm.

The significance of this dynamic compounds in the case of a technology, such as nuclear power, that appears structurally problematic on other grounds (e.g., risk to future generations). In extreme cases, a technology could conceivably cause enough external harm to damage the prospects for replacing it via uncoordinated market decisions, while also causing enough structural harm to damage the prospects for replacing it via a democratic political decision. For instance, the hazards of nuclear power have already contributed to its being managed by relatively undemocratic, technocratic institutions and—if a reactor is ever badly sabotaged or if plutonium or a nuclear bomb is stolen—could conceivably provoke socially extensive, governmental authoritarianism.[26]

Externalities can thus produce a prolonged cascade of dynamic social responses (or pathological nonresponses), including an unceasing stream of new external effects. This cascade effect further discredits the notion that self-regulating markets yield Pareto optima, and in addition challenges the economistic notion that market behavior "reveals" people's preferences. During such cascades—which, given externalities' pervasiveness, may well be the historical norm—people are buffeted by social forces beyond their influence. If a particular cascade ever does

subside, there is no reason to imagine that the resulting social circumstances are generally ones that people would have chosen had they had
a say in the matter.[27]

Moreover, if not politically anticipated or guided, such dynamics
can impair the structural conditions requisite to sustain democratic
politics, including the political capacity to oversee or regulate subsequent market interactions and externalities. For instance, in allowing
public life embodied in vibrant downtown public spaces to be eclipsed
as an external consequence of, say, patronizing shopping malls and
watching television, citizens reduce their democratic capacity either to
guide subsequent local developments or to influence translocal events.[28]

Fourth, the very goods we buy or create for ourselves exhibit
properties and acquire meanings that we cannot anticipate when we
acquire them. Consider Milan Kundera's account of the dense layers of
meaning accumulated by a bowler hat:

> Let us return to [Sabina's] bowler hat:
>   First, it was a vague reminder of a forgotten grandfather. . . .
> Second, it was a memento of her father. . . . Third, it was a prop for
> her love games with Tomas. . . . Fourth, it was a sign of her originality
> . . . . Fifth, now that she was abroad, the hat was a sentimental object
> . . . . The bowler hat was a motif in the musical composition that was
> Sabina's life. It returned again and again, . . . and all the meanings
> flowed through the bowler hat like water through a riverbed.[29]

Alternatively, as individuals age or change psychologically (partly
as a result of their own economic activities), their new self experiences
and judges differently the personal world inherited from their former
self. Thus even their own acquisitions produce externalities (i.e., effects
not mediated by prices in a perfect market) for them, not just for other
people.

If one adds this insight to those preceding, the result, as one can
also confirm by looking about oneself, is that externalities are not merely
pervasive, coercive, and seductive; they also substantially constitute our
entire world. In other words, contrary to the neoclassical economist's
vision of an ideal world in which people get exactly what they pay for
and nothing else,[30] the real world is, and must invariably be, composed
almost entirely of entities and events that individuals didn't buy (e.g.,
the purchases and activities of neighbors, fellow workers, employers,
corporate executives, and so on), that they didn't buy voluntarily (e.g.,
public goods financed by taxes), or that they experience quite differently
than they expected when they bought them.

However, in constituting our world, externalities in particular help constitute our society's basic structure. This should not be entirely surprising; it simply restates one of this book's central theses: "private" or local technological actions can produce enduring public consequence. Hence the dynamic of structuration can be rendered in economic terminology: *people's actions produce superfluous results (externalities) that, in the aggregate, help comprise structures that affect fundamentally who they are, the world they inhabit, and what, individually and politically, they can and cannot accomplish.*[31] Ritual claims regarding the efficiency of competitive markets are oblivious to this deeper reality that the entire human world has become substantially an ongoing, unintended economic by-product.

Opportunity for Strong Democrats, Problem for Economism

An awareness of structuration can help embolden those concerned with basic social arrangements to compete with economists for the right to a hearing concerning the exigencies and ordering of everyday life, including decisions about technology. However, for economism, the same awareness generates serious analytic difficulties. While trying to prescribe conditions under which individuals can satisfy their (ostensibly exogenous) preferences, economists' models do not account for the structurational reality that individuals are, through their economic and other social activities (including involvement with technologies), continuously transforming the conditions that shape their preferences. Even economistic policy prescriptions contribute unintentionally to establishing conditions that influence the content and ordering of preferences.[32]

Hence, instead of facilitating satisfaction of exogenous wants, economism—contrary to its intention or self-understanding—promotes a tail-chasing quest for efficiency in satisfying wants that economic interaction, economic institutions, and economistic advice all help continuously and covertly to generate.

NORMATIVE DEFICIENCIES

Closely related to economism's analytic confusion are both internal and external normative deficiencies. These derive in part from economism's insensitivity to structuration, revealing a deeper sense in which that insensitivity amounts to a serious fault.

Internal Critique

Besides treating preferences as exogenous, economism is committed morally to the view that individuals' preferences are both arbitrary and sacrosanct—that is, neither rationally defensible nor criticizable. This ethical conviction is reflected in economistic arguments about the sanctity of consumer preferences and in the ordinary economistic defense of competitive markets on the grounds that these respect consumer sovereignty. Economistic policy analysts generally agree that a cardinal sin of public institutions or of professional conduct is to infringe people's freedom by interfering with their views of what they want for themselves. As a standard text puts it, "With rare exceptions, we should accept individuals' own judgments as the appropriate indicators of their own welfare."[33]

When juxtaposed with the reality of structuration, this normative commitment generates a fundamental moral inconsistency: insofar as both competitive markets and economistic policy prescriptions contribute structurationally to shaping individuals' preferences, they are violating economism's own moral canon. Economism is thus deficient by its own standards of morality, descriptive adequacy, and logical consistency—and not merely with respect to incidental matters but concerning its core mission (satisfying preferences).

The theory of democratic technology provides, however, an external standpoint from which to advance further normative criticism of economism.

The Foundation of Social Welfare

From the standpoint of economism, a more satisfactory framework for prescribing social policy would presumably have to reconcile the reality of structuration (i.e., the enduring social consequences of individuals' economic and other actions) with the desirability (as economism phrases it) of not infringing individual freedom. Beyond this, neoclassical economists would probably be impressed if one could make headway dealing with one further normative concern of theirs: social welfare. Unlike efficiency, this is a concept on which economists readily concede considerable perplexity. As they formulate it abstractly, social welfare involves appropriately combining economic efficiency with equity (i.e., fairness in the social distribution of economic benefits and costs). However, economists have not been able to decide on a determinate procedure for specifying that "appropriate combination" in practice.[34]

A democratic politics of technology (or, more generally, democratic structuration) addresses these interconnected issues by reworking several key economistic concepts. First, this chapter has already shown that the concept of efficiency in satisfying preferences is implicated deeply in contradiction and confusion. Hence, whatever else may be required to optimize social welfare, it cannot involve efficiency in this sense (or, rather, this non-sense).

The next step builds on the intimate connection, expounded by Kant, between individual freedom and morality. This allows the concept of freedom to be integrated with that of structuration and to play a more decisive role in a revised concept of social welfare. Chapter 3 argued, in effect, that insofar as a society's social structures influence its members' preferences, psychological capacities, and opportunities to act, those structures ought to establish maximum equal chances for people to develop and exercise moral autonomy, including via participation in guiding subsequent structural evolution.

Rather than follow economism in regarding freedom and fairness as distinct norms, the idea of democratic structuration thus incorporates a notion of fairness into the concept of freedom, conceiving of maximum equal freedom, moral autonomy, and mutual respect as the fundamental ground of social welfare. But, the structural conditions of democracy—including a democratic technological order, as sketched in Part II—do not entirely constitute the substance of social welfare. The remaining substance must be decided through democratic deliberation.[35] The resulting specification of social welfare may vary over time and from one polity to the next so long as strong democracy in some reasonable variant is perpetuated. As hoped, this advances beyond economism both by reconciling structuration with freedom and by developing a more operational concept of social welfare.[36]

External Critique

I have argued that economism is confused and morally inconsistent concerning its central mission. However, from the standpoint provided by democratic structuration, economism fares even worse. This external standpoint suggests that aspects of social and economic life that economism fails to discuss (e.g., endogenous development of preferences and structuration) bear profoundly on people's highest order interest in freedom and democracy. That is, the same economic processes that are not satisfying exogenous preferences are nevertheless arbitrarily—and sometimes detrimentally—reworking a society's basic structures.[37]

Consider what this means for economism's familiar evaluative categories and indicators (e.g., economic cost, cost–benefit ratio, rate of return on investment, productivity, and growth). According to economism, these variables each bear directly and meaningfully on social welfare. When, for instance, aggregate economic productivity increases—assuming no related adverse changes in any other economistic variable (including equity)—then social welfare increases along with it. From the viewpoint of democratic structuration, this economistic understanding is false. Not only is economism's own concept of social welfare internally muddled, but none of the preceding economistic variables is determinately linked, in magnitude or even direction of change, with welfare's actual, highest order dimension (i.e., with establishing democratic structures).

For instance, consider a means of increasing productivity that either psychologically or organizationally stunts workers' autonomy (e.g., certain forms of automation) and that produces commodities that, although marketable, are substantively harmful to democratic community. This productivity increase ought to register as a democratic welfare loss. However, economism would count it as a welfare gain and lump it together with all other productivity changes, some of which might indeed be neutral or positive in their structural bearing on democracy.[38] Or consider how economism normally celebrates the enhanced agricultural productivity of hybrid seed as an unequivocal welfare boon. Plants grown from hybrid seed consume extra fertilizer and pesticide and are infertile or produce only inferior seed, causing farm communities to become dependent on petrochemical companies and large commercial seed suppliers. Insofar as both local self-reliance and ecological sustainability thus decrease (contrary to Design Criteria F and H), hybrid seed's actual effect on social welfare is ambiguous.[39]

Combining these external criticisms with the previous internal ones, one must conclude that economism—however worthy its intentions—is more than confused and inconsistent. It also obscures and frustrates our highest order interest in democracy and moral autonomy. (It strains credulity to imagine that all of economism's flaws are there entirely by accident or that economistic methods are so much in vogue solely because their rationale and reasoning appear so compelling. By ignoring moral autonomy, democracy, and structuration, economism serves an obvious ideological function in obscuring power relations and deflecting political scrutiny away from structures and practices that subordinate democracy to the short-run material interests of a privileged minority.[40]) Thus, relative to a conceptual framework that is sensitive to structuration and concerned foremost with establishing democratic structures, shouldn't economism take an unequivocal back seat in public decision making, technological and otherwise?

This question requires amplification. If economism means a democracy-adverse preoccupation with indefensible concepts of efficiency and social welfare, then even the back seat is unwarranted. Why shouldn't economism in this guise simply be discarded? On the other hand, certain conceptually subordinate economistic indicators, such as productivity and growth, are empirically grounded and track real-world events—even if the welfare implications of those events are not simple to decide and certainly not as economism rates them. Thus, in general, there is no fundamental reason not to take such indicators into account in policy deliberations, provided that (1) one wishes to do so, (2) their welfare implications are interpreted cautiously rather than economisticly, and (3) they are taken into account only after attending to alternative policy options' higher priority structural implications for democracy. It is reasonable, for instance, to take economic evaluations of implementation costs into account when choosing among alternative technology policy options that democratic deliberation has already judged to be substantively democratic.[41]

There is only one general circumstance when empirically informed elements of neoclassical economic knowledge ought to share the front seat with considerations that bear structurally on democracy. This is when the real-world phenomena that economism's statistical indicators attempt to represent themselves develop structural implications for democracy. The obvious example is an economy that cannot meet citizens' elementary needs for biological well-being and survival.[42] In such cases, empirically grounded economic knowledge should be taken judiciously into account on a high-priority basis. However, this should occur for no longer than it takes to mitigate the justifying circumstances, and only because here economic conditions translate into a strong identifiable bearing on democracy. Even then, these technical and economic issues should operate merely on a par with other considerations that bear on democracy—never, as today often occurs, preempting them.

In practice, one can anticipate many occasions arising when someone will assert—perhaps sincerely, perhaps only self-servingly—that a contemplated economic impact does bear structurally on democracy. That would be fine; let democratic forums reach their own conclusions in each such case. The idea is not to banish economic considerations from politics, but to rewrite the rules. Economistic analysts, if they wished to be taken seriously, would have to learn to express themselves in terms of morality, democracy, and social structures. Other people, instead of being encouraged to translate their deepest concerns into economistic terminology that can disrespect and distort them, would be empowered to express their convictions as they actually feel them, affirming their dignity as moral agents and citizens.[43]

Of course, democratization needn't invariably prove economically punitive. Many steps toward democracy, while doubtless eliciting opposition when they challenge power, promise economic benefits. Democratic workplaces, for example, are frequently more productive than authoritarian ones.[44] Greater self-reliance can protect a community against costs imposed by disruption in distant markets (e.g., from war or trade embargoes) or by a distant corporation's decision to close a local branch. Measures to enhance ecological sustainability, by reducing inputs or waste, often lower production costs. Communitarian/cooperative technologies can capture economies of scale in consumption and, along with democratic public spaces, provide social satisfactions that make it less necessary for people to make costly investments in compensating private goods, such as home entertainment devices, home and yard maintenance equipment, and vacation homes. If, as is at least conceivable, enhanced social interaction and civic engagement were to lower the incidence of criminal behavior, then police, other criminal justice, and public and private security costs would also decline.

Many of these sorts of economic benefits of democratic technology are exemplified in the case of the Amish, who maintain high employment, accumulate wealth, live relatively sustainably, produce impressive agricultural yields per acre and per unit of energy input, are little affected by vagaries in interest rates or energy prices, and are financially and socially secure (even while refusing to participate in the federal Social Security system).[45]

POLITICAL DEFICIENCIES

The normative framework developed here also differs from economism in terms of the actors that it requires to use it. As applied within the context of pending political decisions, economistic methodologies are often portrayed as value-neutral machines operated impartially by expert practitioners to generate objective social policy analyses or recommendations. Such simplistic portrayals invariably prove misleading or self-serving; even introductory texts in economistic techniques as much as say so:

> Benefit–cost analysis is especially vulnerable to misapplication through carelessness, naiveté, or outright deception. The techniques are potentially dangerous to the extent that they convey an aura of precision and objectivity. Logically they can be no more precise than the assumptions and valuations that they employ.[46]

Without challenging the techniques themselves, such texts urge aspiring analysts to be aware of their biases, use the techniques carefully, and check how results vary with alternative assumptions. Insofar as they constitute only a few sentences interspersed within texts that are otherwise relentlessly upbeat, it is unlikely that the occasional sobering caveat permeates deeply into many students' consciousness. However, that is not the most important point.

Value-Neutral or Antivalue?

A more serious claim, because it obscures the need for democratic deliberation in technological decision making, is that economistic methodologies are value-neutral. Beneath that outward guise, these methodologies conceal one very important value judgment: economism is committed to the view that all values are fungible (i.e., they can all be traded off against one another).

The judgment that all values are fungible is a value judgment. It is therefore inconsistent with economism's commitment to value-neutrality. Once this value judgment has been pointed out, economism cannot possibly defend it without violating its allied conviction that all value judgments are ultimately arbitrary and rationally indefensible. However, rather than belabor these internal inconsistencies, let us examine economism's latent value commitment from the external standpoint of democratic morality.

One foundational idea underlying cost–benefit analysis and allied techniques is unobjectionable: the simple notion that in making decisions, one should try to take into account their good and bad aspects. But the moment one steps beyond that truism, economism slips into difficulty. How should one decide what to count as a good and what as a bad? Economism offers no unambiguous answer.[47]

In contrast, a neo-Kantian perspective argues that moral autonomy is a precondition for deciding what is good and bad, and therefore autonomy and its necessary conditions are highest order goods. In other words, as supreme goods, autonomy and democracy cannot reasonably be compromised in order to obtain any other putative good. Thus, in contrast with economism, within a neo-Kantian perspective values are not all fungible.

Thus one complaint against economistic methodologies is that they are inadequately structured and morally unprincipled. Despite aspiring to respect the sovereignty of individuals' wishes, economism reduces the ultimate meanings in people's lives to a mash of arbitrary and indefensible preference—a goulash in which, for instance, freedom, community, religion, sacred places, love, and respect are forced to compete on a par with chewing gum, cheeseburgers, underarm deodorant, and other goods whose value, unacknowledged by economism, actually depends on democratic background structures. To put it another way, would it have been reasonable for, say, the U.S. Founding Fathers to have subjected the proposed Bill of Rights to a cost–benefit test? Would we be satisfied if they had rejected freedom of speech on the grounds that the supposed benefits could not hold a candle against the obvious economic costs? By regarding all values as equally irrational, economism exhibits a peculiar species of value-neutrality that—instead of respecting value, purpose, and meaning—tends via homogenization to obliterate them.

Democracy could, of course, be included as a variable in cost–benefit analysis. But cost–benefit analysis does not entail doing that (and in practice democracy is almost universally omitted); it would allow democracy to be traded off against any other value, no matter how incommensurate[48]; it offers little guidance on how to define freedom and democracy if they were included; and it is oblivious to the empirical conditions of democracy and to their source in structuration.

Moreover, when it comes to such elementary and ubiquitous matters as how to decide which alternatives to evaluate; how to neutrally derive a shadow price when (as is always the case) there are manifold and multidimensioned externalities, often producing combined synthetic effects; and how to weigh and trade off against one another qualitatively disparate phenomena (some of which are held sacred and so ought not to be traded off against mundane values), economistic methodologies are inept or silent.[49] In practice, analysts end up doing whatever they like, perhaps quite unaware of the subjective nature of the enterprise.

The character of these techniques is thus contrary to the claim, also advanced in their support, that they force analysts to clarify and display their values. These techniques do not promote explicit display or accurate labeling of the countless subjective or subconscious judgments of discrimination, selection, and valuation that are essential to even their most trivial uses.[50] For instance, analysts who routinely omit all structural political consequences rarely know it or say so if they do.

If existing analytic techniques are imperfect, why not correct them? The task is impossible. Attempting in good faith to remedy these techniques, one transforms them beyond recognition. What results is neither a machine nor any longer a pseudomachine but instead a contestable language of democracy (see below).

Actors

Whether employed "correctly" or "abusively," economistic methodologies in effect disenfranchise citizens who lack technical credentials.[51] Would a democratic politics of technology and its supporting contestable design criteria inadvertently do the same? The approach presented here militates against that result: it does not pretend to be a value-neutral machine. Instead, its philosophical structure imposes cognitive demands on users that, in turn, translate into democratically favorable procedural requirements.

This approach is not value-neutral, because it places freedom and democracy prior to other values and because it recognizes that, within a context of democracy, values can be discussed openly and reasonably. It is not a machine, because there can in principle be no such thing as a fully specified, context-invariant conceptual method for assisting democratic structuration. Instead, the conditions of democracy (as reflected, for instance, in Part II's design criteria) interpenetrate the fabric of daily social existence, are individually contestable, sometimes mutually compete, and can evolve.

Accordingly—now moving on to cognitive demands—to identify considerations that bear on democracy and to bring them into a harmony that effectively embodies freedom requires creative and judicious sorting, weighing, and disputation on a contextual basis. These are tasks that, while they can usefully be informed by democratic theory, cannot be reduced to prespecified, machinelike judgmental operations. Moreover, part of the context that democratic deliberations must take into account includes citizens' own diverse conceptions of their collective situation, their common good, and their preferred options for furthering the common good (including, preeminently, alternative strongly democratic policies and institutional arrangements).

All these activities intermingle judgments of fact with judgments of value. Thus a strong democratic framework affords no easy recourse to the ostensible objectivity of expert testimony on matters of pure fact.[52] Rather, one seeks impartiality by ensuring representation of all competing viewpoints in deliberations that are as open and egalitarian as possible.

Let's turn next to the political implications of the preceding cognitive requirements. First, reliance on creativity militates against expertise-based restrictions on the legitimate participants in political deliberation: the more actors, the more potential for novel perceptions and ideas. In the words of a Zen master, "In the beginner's mind there are many possibilities, in the expert's there are few."[53] This provides some initial grounds for permitting, and even encouraging, citizen participation. These grounds are then extended by the need to acquire contextual knowledge. This demands actively soliciting citizens' involvement, as took place during Canada's Berger Inquiry.

Finally, reliance on reflective judgment informed by egalitarianly balanced, competing viewpoints, suggests more. Not only should citizens be encouraged to participate, but—relative to experts—they should be sovereign. No one other than a collectively self-educating citizenry can, acting through complementary representative and participatory institutions, competently undertake the deliberations necessary to promote and sustain democracy. No one but citizens themselves are positioned to discover the facts, contest the criteria, and implement the decisions necessary to express their own freedom. Procedure and outcome count equally here, because the very process by which citizens make their judgments develops and expresses their freedom and, as its results become embodied in action and structure, further supports their freedom.

In short, the concept of democratic structuration, including a democratic politics of technology with contestable design criteria, does not amount to another pseudo value-neutral, quasi-cookbook methodology that can be appropriated by the next generation of professionally credentialed policy analysts. For this reason especially, one can expect economistic analysts to find it difficult to recognize the approach presented here as an alternative, much less a potentially superior one. Even if a minority are willing to contemplate significant alternatives to their methodologies, probably few indeed are willing to contemplate one that would reduce their privileged position in economics and politics. This does not mean that the argument presented here is flawless; certainly economists and policy analysts have every right to criticize it. However, one must always be wary of criticism that has as its real function or intent preserving illegitimate power relations.

In place of a formulaic methodology, I propose a fluid and contestable language of democracy. Fluency in such a language may not come effortlessly. However, any citizen can learn to critically appropriate the rudiments and contribute to producing it.[54] Because the language respects people's supreme interest in freedom, using it ought to enhance their self-worth and dignity. Citizens may also find some previously slighted moral intuitions clarified and affirmed. If anything, moreover, it is probably easier for laypeople to master democratic discourse and practice than for professionals entrenched in economistic modes of thought to do so. For instance, Sagoff describes economists perplexed by citizens who obstinately refused to cooperate in quantifying matters of deep personal or social concern, while nonetheless readily supplying unsolicited moral and political judgments.[55]

Economism is a language in terms of which citizens never truly win.

Even if they manage to win an immediate political battle, they have nonetheless—by failing to respect one another's moral capacities and deepest values, and by tacitly legitimating antidemocratic political processes and sociotechnological structures—lost more than they have gained. In contrast, a democratic politics of technology promises citizens a win even when they lose. Even if citizens lose an immediate political battle, provided that they have expressed themselves in a nondebased language of democracy, they affirm the dignity of all and help create a political climate conducive to that critical scrutiny of a society's basic structures that is democratically essential.

CONCLUSION

Despite its embarrassing and intractable deficiencies, economism reigns unchallenged in contemporary technological decision making. However, by offering greater descriptive realism, coherence, and internal consistency, and a defensible approach to determining social welfare, the concept of a democratic politics of technology seeks to best economism on economism's own turf, while also proposing independent grounds for being judged superior.

To the extent that this chapter's arguments prove able to withstand fair-minded critical scrutiny, they should furthermore qualify as one reason to favor strong democracy—in the specific variant elaborated here—over thin democracy.[56] That is, this chapter's critique derives specifically from neo-Kantian morality and from structuration's dialectical logic and descriptive and analytic insights. These are at odds with the philosophical premises—for example, egoism, utilitarian ethics, and nondialectical logic—underlying and informing thin democratic theory and practice.[57] Consequently, thin democratic theory cannot share in any credit due strong democracy for besting economism. Indeed, insofar as thin democracy shares philosophical premises with neoclassical economics, this chapter's demonstration of inadequacies in those premises could conceivably induce some unsettling aftershocks in thin democracy's foundations.

On the other hand, in criticizing economistic theory and methods, I have not argued for eradicating competitive economic markets. The conventional economic justification for markets on efficiency grounds is false. Markets do not satisfy exogenous preferences, but do clandestinely and often adversely transform social structures. Conventional markets also nurture egoism, not moral development or citizenship, and indeed often punish nonegoism. They permit grossly undemocratic concentrations of wealth and power and even reward producers for

discovering ways of sundering self-reliant communities into aggrega-
tions of dependent atomized consumers.

However, competitive markets do clear. They can also, under the
right circumstances, constitute a domain of personal initiative and an
important means of coordinating social interaction without overpoliti-
cizing everyday life.[58] Subjecting every action, including ones obviously
innocuous, to political scrutiny would pervert democracy into an obsession with control for its own sake and would waste time and effort.

Markets are incompetent to function in place of strong democratic
politics or to be allowed tacitly to determine social policy or structural
transformations. However, refashioned and tamed, they can serve as
democratic policy instruments. Aside from democratizing corporations,
redressing gross disparities in wealth, and seeking greater local economic
self-reliance, this entails struggling toward democratic procedures for
overseeing and, as warranted, regulating market dynamics.[59] Citizens
must be able to ensure that the aggregate structural consequences of
economic interaction, taking into account static and dynamic external
effects, remain consistent with democracy and the common good.

Chapter 11

"EVERYONE CONTRIBUTES"
Participation in Research, Development, and Design

> We saw the big printing companies
> and newspapers doing their own
> research. They were giving money
> to . . . develop new equipment that
> we felt would undermine our
> traditional labor agreements. So we
> thought it was important for us to
> . . . support technology that would
> lead to the kind of skills and
> working conditions that *we* were
> interested in.
> —*Gunnar Kokaas, secretary of the
> Norwegian Graphics Workers Union*[1]

Suppose citizens achieve extensive opportunities to evaluate and choose technologies, but the range of available choice is quite narrow? Or suppose, as today may often be the case, the range of choice includes only relatively nondemocratic technologies? Democratic participation in technological politics could amount to little more than a hollow exercise or sham.

In the language of political science, technological research, development, and design thus emerge as the agenda-setting phase in technological politics.[2] If it is vital that citizens be empowered to help shape legislative or electoral agendas, it is likewise vital that they have extensive opportunities to participate in technological research and design. Only in this way can people be assured that when the time comes to choose among alternative technologies, the menu of choice will afford an adequate opportunity to express their personal aspirations, view of the common good, and understanding of democracy's necessary conditions.

Broadened participation in the research, development, and design (RD&D) process will tend to issue in a wider range of designs for several reasons. First, a larger number and more diverse range of participants increases the chance that someone will come up with a creative insight. Second, it ensures that a more diverse range of prior social needs, concerns, and experiences will be reflected in the design process. Today, in contrast, relatively homogeneous groups, mostly consisting of professional white males, dominate design processes and indeed share a latent interest in suppressing innovations that might subvert their privileged social position.[3] Third, broadened participation can provide enhanced opportunities for fruitful cross-fertilization of ideas from one domain of social and technological experience to another. In the words of Professor Michael Porter of Harvard Business School, "Innovators are often outsiders."[4]

Competitive economic markets already permit some opportunity for consumers and stockholders to influence corporate design decisions. However, markets alone are democratically deficient in numerous respects.[5] For instance, as a solitary consumer, a person can do nothing to influence the aggregate structural results of the totality of consumer and producer decisions. Consumers also have little power to influence decisions concerning production, infrastructural, and military technologies. Consumers cannot indicate their preference for goods and services that are not available on the market. Markets and firms are much more responsive to individual wants than to collective needs (e.g., for public goods or for communitarian/cooperative technologies). Markets are disproportionately attentive to the wants of the ultrawealthy, while not at all responsive to the urgent democratic needs of the economically deprived.

This reality establishes sufficient grounds for democratizing the processes of technological RD&D. However, there are further grounds.

SELF-ACTUALIZATION

RD&D activities can provide rewarding opportunities for creativity and self-actualization. Successful inventors sometimes mention moments of insight in which they felt ecstatically transported.[6] One member of a participatory, cohousing design project recalls "learning how to make critical decisions with others and gaining an immense amount of self-confidence."[7] A solitary inventor, who was struggling to juggle her competing responsibilities as a mother, nonetheless

found that I was engrossed in . . . [inventing] till all hours of the morning. . . . But, I felt I had to, I had to fulfill what was inside of me.[8]

A 12-year-old boy, engaged in a year-long collaborative school project to redesign his own city, wrote:

> When you build a city, you see how important your job is and everyone contributes. I guess I realized I'm not just a nothing person. I'm a little important person.[9]

Insofar as participatory design results in technologies or services that go on to help constitute the agenda for a democratic politics of technology, then RD&D can also embody that special dignity that attends helping consciously to evolve one's society's structural form.

However, RD&D processes can also be organized to be crimped and growth-impairing. In the recollection of a former machinist:

> I began . . . meeting many men whose lives revolve around the design and development of machines. I remember a dullness about them, as though the gray of the steel had entered their souls. . . . I too felt a certain lifelessness growing in me.[10]

That much modern RD&D occurs under circumstances that permit little creativity is also suggested, albeit inadvertently, by technological apologist Samuel Florman's 1976 book, entitled *The Existential Pleasures of Engineering*. The volume celebrates the virtues of modern engineering, but when it tries specifically to demonstrate that engineering can be creative and fun, every example concerns artists or scientists. In other words, it was apparently not easy to locate instances of professional engineers who describe their work in uplifting terms.[11]

In short, concerted efforts are needed to redesign RD&D settings so that they indeed provide opportunities for creativity and self-expression. The experience of a British industrial designer is telling. Employed by a traditional aerospace corporation, he suddenly had a chance to collaborate nonhierarchically on a socially useful project of his own choice, while interacting directly with the intended beneficiaries of his work:

> [Our trade union] members . . . visited a centre for children with spina bifida and were horrified to see that the only way the children could propel themselves about was by crawling on the floor, so they designed a vehicle which subsequently became known as hobcart. . . . Mike Parry Evans, its designer, said that it was one of the most enriching experiences of his life when he delivered the hobcart to a child and saw the pleasure on the child's face. It meant more to him, he said, than all the design activity he had been involved in up till then.[12]

POLITICAL EDUCATION

A final set of reasons for democratizing RD&D involves political education. First, participation can be a means of discovering the general fact, otherwise often obscured, of social contingency in technological endeavor.[13] This is a crucial step in becoming motivated to function as a competent participant in technological politics. Second, involvement in design, especially within socially open and democratically organized settings, tends to elicit critical reflection on social circumstances and needs. This is vital to democratic politics in general, and particularly with respect to technology. Third, from a societal perspective, more diverse participation in RD&D can help ensure—via broadening of the collective perceptual field—that a wider range of nonfocal technological consequences will be articulated within design processes as well as in subsequent choice and governance procedures.

Fourth, and perhaps most importantly, broadened popular involvement in design would provide crucial opportunities to publicize technological developments at the stage—prior to the evolution of strong institutional commitment, enduring material embodiment, uncritical habituation, or social dependence—when there is by far the greatest potential to influence their course.[14] Contemporary technological politics (e.g., concerning nuclear power or advanced weapons) typically occurs years or even decades after crucial design decisions have already been made. At that time the only feasible choice is generally a heroic yes or no. However, "yes" risks leaving opponents bitterly disillusioned concerning both the outcome and the political process leading to it,[15] whereas achieving a "no" requires tremendous political mobilization and risks jeopardizing jobs and enormous investments of developmental money, time, and talent. For instance, when Wisconsin enacted a temporary ban on farmers' use of a genetically engineered growth hormone (proposed as a stimulant to cows' milk production), the Monsanto Company stood in danger of losing 300 million dollars in prior research and development costs.[16]

In contrast, early public involvement and publicity allows flexible, socially responsible, research and design modifications all along the way. Instead of a crude yes or no, accommodative adaptations are possible. This promises greater social acceptance of the RD&D and political processes, greater social satisfaction with their substantive results, and hence a less adversarial, more fair and economical path of social structural evolution. For instance, Peter Johnson, former head of the Bonneville Power Administration, reports in a 1993 *Harvard Business Review* article that

> public involvement is a tool that today's managers in both public and private institutions must understand. . . . We found that by inviting

the public to participate in our decision-making process, our adversaries helped us make better decisions.[17]

Johnson notes, for example, that one extended participatory process—which helped set electricity rates affecting the future of the Northwest aluminum industry—resulted in a decision that served the public good, while also enabling Bonneville to earn more than $200 million that it would otherwise not have received. Others argue that earlier, empowered participation holds promise as the most fair and effective alternative to the NIMBYism (Not-in-My-Back-Yard syndrome) that has paralyzed many recent attempts to site new industrial and waste-management facilities in the United States.[18]

Moreover, to the extent that the pace of technological change has recently accelerated, arguments for participation in RD&D gain added weight and urgency. When technological change proceeds at a leisurely tempo, it can be possible to learn and to make some adjustments retrospectively. Now, however, often by the time a technology has been deployed and experienced, replacement technologies are already designed and entering commercialization. For instance, linotype technology for typesetting underwent no basic changes for 100 years. But in the last 20 years, since linotype machines began to be replaced, there have been several successive, fundamental shifts in the technology and social organization of commercial typesetting. Under these circumstances, retrospective adjustment is not an option.[19]

Finally, each proposed or realized design can help enlarge people's awareness of the specific range of functions, effects, and meanings associated with other designs. For instance, the contrast between digital and analog wristwatches has led me to experience the numeric displays of the former as easy to read but also unnecessarily precise, tending to elicit compulsiveness and a disjointed sense of time. I now consciously experience the latter—with their circular faces and cyclical displays—as connecting me, in some admittedly imperfect way, with rhythms in nature. Were citizens gradually to become aware that, whether or not it is materially realized, a publicized design alternative can thus prove educative, this would further enhance the dignity of participation in design processes. Good in itself, that would in turn increase personal incentives to participate.

DESIGN COMPETENCY

Are nonexpert citizens capable of becoming competent to participate in design? The conventional wisdom says no. As Green puts it, "We do not really expect a committee of workers . . . to sit around a table inventing microchips or the cathode tube together."[20] But, as often proves to be true, the conventional wisdom seems ill-informed.

Part of the reason is that research, development, and design are complex social processes, implicitly embodying many options for different ways of participating. Notwithstanding the mythology of the solitary inventor, RD&D is intrinsically a social and political activity, one involving incessant collaboration, disputation, and compromise.[21] Within contemporary firms or government laboratories the range of participants already includes, indeed is often dominated by, many people not technically credentialed (e.g., entrepreneurs, bureaucrats and managers, planning and marketing professionals, financiers, lawyers, accountants, policy analysts, and congressional staff). Moreover, the actual process of design is by no means restricted to the activity of design professionals. People design when they decorate a room, create and serve a meal, or choose how to dress[22]; workers often contribute to design in ways that go unacknowledged[23]; communities engage in design when they formulate development plans or zoning laws; and legislatures, courts, and administrative agencies participate in design when they promulgate laws or regulations demarcating realms of permissible and impermissible technological endeavor. Modern high-productivity manufacturing methods, such as Japanese just-in-time production, depend on harnessing the creativity of anyone and everyone—including managers, staff, workers, suppliers, and customers—in an unceasing quest for incremental improvements in manufacturing processes and products.[24] Even high-tech innovations are often less math- or science-intensive than is commonly supposed.[25]

For these reasons one can envision many different ways in which people might become involved in RD&D. That range would increase if, as should come about, there were broader social opportunities to develop and extend one's design-relevant competencies. Opportunities for participatory RD&D will also tend naturally to expand as various other steps toward technological democratization are implemented, such as citizen sabbaticals, flexible careers, democratically managed workplaces, and more diversified, self-reliant local economies.

A series of case studies of participatory RD&D in diverse social domains can shed further light on these issues.

Workers and Citizens Designing Socially Responsible Technologies

During the late 1960s, Britain's Lucas Aerospace company sought to lower costs by closing some factories and laying off workers. In response, the workers, who were organized in a companywide combination of unions that included engineers, proposed to maintain employment by converting the factories to manufacture a new range of products. Soliciting ideas and assistance from universities and local communities, but relying substantially on the latent creativity of all ranks of Lucas workers, the union developed an alternative corporate plan that satisfied several new criteria. These encompassed a commitment to producing more socially useful goods (rather than, say, military jet engines), while relying on more democratically organized and creative work processes. The Lucas Aerospace Workers' Plan eventually included 150 product ideas, accompanied by economic and engineering analysis and some working prototypes. The ideas ranged from inexpensive medical devices to solar energy hardware, fuel-efficient vehicles, multifuel/multiuse electric generators adapted to third world village needs, and worker-controlled robotic devices designed for operation in hazardous environments.

The proximate tactical results were not encouraging. The company toyed with several of the new products but rejected the basic concept of worker and community participation in design and strategic planning. Lucas also managed—after Margaret Thatcher's Conservative party came to power—to fire union leaders. However, even these events were not without value: they disclosed some of the political–economic barriers to technological democratization and provided a springboard for follow-on endeavors.

Through publications, lectures, radio interviews, television documentaries, and incorporation into the televised curriculum of a popular Open University continuing-education course on social control of technology, the Lucas saga spread widely. Similar alternative corporate plans have been developed at other British companies and in other nations. Collaborating with North East London Polytechnic, Lucas workers designed a fuel-efficient bus able to travel on both roads and rail, hoping thereby to demonstrate a promising transportation technology while carrying to local communities a multimedia presentation of the Lucas story. The purpose was to help stimulate further political and technological collaboration between local communities and socially minded workers and engineers.[26]

Meanwhile a former leader in the Lucas union, Mike Cooley, went on to become director of the Greater London Enterprise Board's (GLEB's) Technology Division, which set up a series of "Technology Networks." Each Technology Network had a research and advisory staff plus machine tools, related shop equipment, and access to nearby college libraries and technical faculty. Individual citizens, nonprofit groups, and aspiring businesses that met one or more social criteria (e.g., hiring and promoting women or disadvantaged minorities, cooperative management structure, worker or cooperative ownership, human-centered or environmentally responsible production methods, or a commitment to produce goods or services responsive to unmet social needs) received free help from the Technology Networks in developing and marketing socially responsible technologies. GLEB also began offering startup subsidies to small businesses meeting the same list of social criteria. Over 200 enterprises were initiated, with the Technology Networks functioning as a participatory resource for their research and development needs.[27]

The Technology Networks were an important success insofar as laypeople and businesses produced plausible ideas they wanted to develop and created sound working prototypes. However, there were also problems. The Technology Networks were popular and reasonably well funded through local taxes, but they lacked the resources and business sophistication required to compete with large established firms in moving an idea from prototype to production and use.[28] Moreover after several years, Prime Minister Thatcher, who opposed the GLEB's mission and resented its popularity, was able to curtail its public funding. The Technology Networks, aware of their shortcomings, thus lost their major source of financial support before they had a chance to work on solutions.

Workplace Technology

During the 1980s in Scandinavia, unionized newspaper graphics workers—in collaboration with sympathetic university researchers, a Swedish government laboratory, and a state-owned publishing company—invented a kind of computer software unique in its day. Instead of following prevailing trends toward routinized or mechanized newspaper layout, this software incorporated some of the capabilities later embodied in desktop publishing programs, enabling graphic artists and printers to exercise considerable creativity in page design. Known as UTOPIA, this project demonstrated how broadened participation in RD&D can lead to a design innovation that, in turn, supports one condition of democracy: creative work.

UTOPIA was less ambitious in overall scope than the Lucas Aerospace workers' attempt; however, UTOPIA demonstrated that workers could go beyond developing prototypes, contributing in this case to a marketed product. The resulting software was nonetheless not a commercial success; researchers attribute this failure to inadequate financial resources on the part of the company responsible for marketing.

This instance of collaboration between workers and technical experts—limited initially to a single technology within a single industry—occurred under unusually favorable social and political conditions. Sweden's workforce is 85 percent unionized, and the nation's pro-labor Social Democratic party had held power during most of the preceding half century, providing a supportive legislative context.[29]

Participatory design by workers has gradually become something of a social movement in Scandinavia, and there are a burgeoning number of documented case studies. Many of these reveal institutional obstacles or resistance of various sorts. But they are quite uniform in finding that, under appropriate circumstances, workers are more than capable of participating effectively in designing workplace technologies.[30] From the standpoint of a fully developed democratic politics of technology, there are some other limitations to the Scandinavian efforts. In particular, although admirably attentive to the effects of alternative designs on workers or other users, these projects have not been comparably alert to the spillover effects on nonusers, including products' structural social consequences either alone or in combination with other technologies.

Participatory Architecture

Compared with the relative paucity of examples of participatory design of machinery or appliances, there is a rich history of citizen participation in architectural design and urban planning. The range of stories is extremely diverse, including cases in which it proved difficult to motivate participation or in which the design outcome was little different from what professionals working alone might have devised. But in other cases highly novel designs resulted, sometimes evoking resistance from powerful social interests and institutions.[31]

For instance, participatory design by future residents has proven essential to cohousing projects.[32] Another example is the "Zone Sociale" at the Catholic University of Louvain Medical School in Brussels. In 1969, medical students insisted that new university housing mitigate the alienating experience of massive hospital architecture. Architect Lucien Kroll established an open-ended, participatory design process that elicited intricate organic forms (e.g., support pillars that looked like gnarled tree trunks), richly diverse patterns of social interaction (e.g., a nursery school situated near administrative offices and a bar), and a dense network of pedestrian paths, gardens, and public spaces. Walls and floors of dwellings were movable so that students could design and alter their own living spaces. Kroll also provided construction workers with design principles and constraints rather than finalized blueprints, and he encouraged them to generate sculptural forms. The workers did, sometimes returning on Sundays with their families to show off their art. Initially baffled by the level of spontaneity and playfulness, the project's structural engineers gradually adapted themselves to the diversity of competent participants. Everything seems to have proceeded splendidly for some years until the university's administrators became alarmed at the extent to which they could not control the process. When the students were away on vacation, they fired Kroll and halted further construction.[33]

Medical Technology and Barrier-Free Design

During the past 25 years there has been substantial innovation in designing barrier-free equipment, buildings, and public spaces responsive to the needs of people with physical disabilities.[34] Much of the impetus has come from disabled citizens who organized themselves to assert their needs or who helped invent or evaluate design solutions.[35] For example, prototypes of the Kurzweil Reading Machine, which uses computer voice-synthesis to read typed text aloud, were tested by over 150 blind users. In an 18-month period these users made more than 100 design recommendations, many of which were incorporated into later versions of the device.[36]

Recently, AIDS patients and activists have played an influential role in challenging the manner in which new drug treatments are devised, tested, and approved. Concerned with hastening the delivery of experimental medications to critically ill patients, they have helped initiate a new genre of community-based trials, involved patients in trial design, and pressed for applying different statistical assessment methods. Together with sympathetic doctors and statisticians, they have proposed

> changes that cut to the very core of how clinical trials are designed [and that may] institute a process that could go well beyond the treatment of AIDS, altering the way medical research deals with other life-threatening diseases.[37]

Feminist Design

What would happen if women played a greater part in RD&D? One answer comes from feminists who have long been critical of housing designs and urban layouts that enforce social isolation and arduous, unpaid domestic labor on women. As more and more households deviate from the norm of mother and children supported financially by a working husband, such criticism has sharpened. One historical response has been to suggest that if women were more actively engaged in design, they might promote more shared neighborhood facilities (such as daycare, laundries, or kitchens) or they might locate homes, workplaces, stores, and public facilities more closely together. Realized examples have been constructed in London, Stockholm, and Providence, Rhode Island.[38]

Another approach has been pioneered by artist and former overworked-mother Frances GABe, who devoted several decades to inventing a self-cleaning house. "In GABe's house, dishes are washed in the cupboard, clothes are cleaned in the closets, and the rest of the house sparkles after a humid misting and blow dry!"[39] Other feminists have established women's computer networks or sought alternatives to unfulfilling female office work, to urban transportation networks that are insensitive to women's needs, and to new reproductive technologies that divorce women from control of their bodies.[40] An explicit feminist complaint against current reproductive technologies—such as infertility treatments, surrogate mothering, hysterectomy, and abortion—is that women have played little part in guiding medical RD&D agendas, the results of which impose on women agonizing moral dilemmas that might otherwise have been averted or structured differently.[41]

Third World Design

There are numerous instructive cases of lay participation in designing technologies adapted to traditional or developing communities. For example, a Guatemalan appropriate technology station was approached in 1976 by village women complaining of excess wood consumption and smoke from their traditional, open-fire cooking methods. The researchers developed a clever hardened-mud cookstove, called the "Lorena," that burned significantly less wood. Village women played a crucial role in evaluating prototypes—insisting, for example, that the original floor-level design be raised off the ground. The resulting design, later widely adopted, was made of inexpensive, locally available materials and lent itself to user self-construction and redesign in conformance with local conditions.[42]

Another third world example reveals a common sort of obstacle to participatory design. In the mid-1970s, the remote Cree Indian community of Big Trout Lake, Canada, was struggling to cope with sewage-contaminated water. A Canadian government agency proposed an expensive sewage treatment plant that would have served only the small non-Indian segment of the community. The Cree objected and hired a participatory research facilitator and their own technical consultants. Collaborating closely, the tribe and its consultants developed an array of alternative, less expensive waste treatment solutions able to serve the entire community. But rather than welcome this community involvement, the Canadian government reacted defensively and the project stalled.[43]

Designing with Children

In the 1960s, city planner Mayer Spivack helped some Boston children restore a neglected playground. Together, they designed playground equipment that could be built—and continuously redesigned and rebuilt—by kids themselves, using industrial surplus and other scavenged materials. The children quickly exhibited boundless exuberance in fashioning their own, ever-changing, public play space. However, local adults, who were not significantly involved in the process, objected to the youths' makeshift aesthetic and insisted that the city government demolish the playground.[44]

The story of American architect Bob Leathers suggests the possibility of more positive outcomes. Since 1970 he has helped hundreds of communities design, finance, and construct their own playgrounds, using a process partly derivative of an old-fashioned barn raising. Children—preschoolers and up—play a significant part in all phases, including design. Both the social process and the new playgrounds themselves have tended to engender a heightened experience of convivial local community.[45] Educational pioneer Doreen Nelson has likewise spent more than two decades involving children, from kindergarten age on up, in year-long exercises in comprehensive, participatory urban design.[46]

POLITICAL IMPLICATIONS

The issue of laypeople's competence to participate in RD&D seems to pose no insurmountable problems. Laypeople can certainly help identify social needs and concerns and thus contribute to developing functional specifications for new technologies, as they did in most of the preceding cases. In addition, their deep knowledge of local cultural and environmental circumstances can prove vital to the RD&D process—as in developing the Lorena stove or Big Trout Lake's sewage treatment alternatives. Finally, when a design process is effectively structured, laypeople can potentially become full partners in developing design solutions, as in UTOPIA and London's Technology Networks. If children can help design playgrounds and even entire cities, surely there is hope for the rest of us.

Second, participatory RD&D tends, as predicted, to issue in a broader array of design alternatives. However, if the preceding examples are representative, there may also be an incipient tendency for participation to encourage designs that are substantively more democratic. For instance, cohousing, Brussel's Zone Sociale, and Bob Leathers's playgrounds each help establish communitarian/cooperativism and new public spaces, thus satisfying democratic Design Criteria A and G (see Figure 5-1). The Lorena stove and the Zone Sociale's movable walls and floors exhibit unusual design flexibility (satisfying Criterion I). The Lucas workers' energy and transportation technology proposals, the Lorena stove, and the Cree sewage treatment alternatives promoted greater ecological sustainability and self-reliance (Criteria F and H). The Zone Sociale, the Lucas Workers' Alternative Plan, and UTOPIA created opportunities for more creative work (Criterion B). Feminist design, AIDS activism, and barrier-free design seek more equalized social power relations (Criterion D).

These results surpass my own prior expectations for participatory RD&D. I had not thought that, in the absence of explicit reliance on democratic design criteria, a participatively broadened menu of technological choice would necessarily prove substantively more democratic. But that seems to be the case, perhaps suggesting greater latent social readiness to support technological democratization than one might otherwise have hoped and also indicating that participating laypeople have already—quite without the assistance of formal democratic theory—implicitly begun to develop and apply democratic design criteria. Technological democratization does not depend on this fortuitous result, but it is certainly welcome.

Third, successful participatory design sometimes requires a conscious effort to cultivate mutual respect and trust among collaborating laypeople and technical experts. In some cases, formal participation facilitators—who need not necessarily be technically trained themselves—will be required. The Cree of Big Trout Lake hired such a facilitator. A supporting program of education for participation may also prove helpful, although its curriculum might need to focus as much on politics, organizational skills, financial savvy, or nurturing self-confidence as on technical know-how. As Ward remarks, "Education for participation . . . is not education about aesthetics . . . , it is education about power."[47]

Laypeople aren't the only ones with much to learn. As in the case of the structural engineers who worked on Brussels' Zone Sociale, technical experts may need assistance in learning how to learn from and with laypeople.[48] Moreover, technical experts are not expert in the specific paramount problem of designing democratic technologies. Indeed, both their training and professional interest may place them at some disadvantage, relative to nonexpert citizens, in being concerned or competent to seek democratic designs.[49] There is, for example, ample evidence of midlevel managers, many of whom are scientists or engineers, resisting technologies that facilitate workplace democratization, even when labor productivity demonstrably improves.[50]

Fourth, although it is morally important to include and empower the most disadvantaged social groups,[51] morality and prudence alike suggest that others not be excluded. Witness the opposition engendered by activist Spivack's oversight in not including neighborhood adults in his Boston playground design process.[52] A tactical exception to the general rule of inclusiveness may, however, be warranted when full representation of traditionally hyperempowered social groups would stifle wider popular involvement or efficacy. It is possible, for example, that had university administrators been more involved in planning Brussel's Zone Sociale, the project's most exciting features might never have left the drafting table.

Fifth, today the vast majority of RD&D occurs within three institutional settings: corporations, government laboratories, and universities. In contrast, many examples of participatory design have occurred in some other setting—sometimes with assistance from iconoclasts employed within major institutions but generally on the social periphery where financial and other resources are scarce.[53] Even the Swedish UTOPIA project had to struggle to receive government support and then restricted itself to designing computer software because funding was insufficient to undertake complementary hardware RD&D. Indeed, as we have seen, many participatory design exercises have encountered outright opposition from powerful institutions—opposition engendered not because the exercises were failing, but because they were succeeding. The absence of more examples of participatory RD&D would seem to have much less to do with issues of layperson competency than it does with economic, political, and cultural resistance.[54]

It follows that if participatory design of democratic technologies is ever to become normal social practice, then political pressure must be developed for diverting economic resources to it or else for incorporating it into the core social institutions that today undertake most RD&D. Otherwise control of the technological order will remain substantially in elite hands. (That is a key criticism that has been advanced against the strategic choice of some appropriate technologists to work unobtrusively on the social margins, while dominant institutions go about their accustomed business.[55]) In the United States, universities or divisions of government research facilities not engaged in classified military research may be the easiest places to initiate democratic involvement in RD&D, because they receive public subsidies and must, in principle, serve the public interest. Indeed, industry has increasingly won the privilege of participating extensively in both university and government laboratory RD&D efforts. What conceivable argument could there be for denying citizens and community groups the same, if not greater, rights? If the answer is that industry is able to pay its own way, then a plausible rejoinder is that citizens' taxes already foot much of the bill for university and government RD&D.[56]

Participation by workers—rather than by outside citizens—in corporate RD&D is also a realistic near-term target, and experiments with it are already becoming more widespread. In contrast, one can expect that insinuating general citizen participation into either military or corporate RD&D will prove more challenging. Among other things, it would generally require some relaxation of the laws governing military and trade secrecy or a corresponding strengthening of the laws governing freedom of information and expression.[57]

None of these suggested reforms is politically trivial. On the other hand, advocates of participatory RD&D have often elected to state their case entirely in terms of current nonparticipants' material interests. Others cite the contribution participation can make to improved worker productivity or to better design solutions. These arguments are reasonable and sometimes effective. However, they neglect the specific moral argument that the opportunity to participate in RD&D is a matter of moral right—essential to individual moral freedom, to human dignity, to democratic self-governance, and to generating technologies more compatible with democracy. Moral arguments, when compelling, are right to make not only because they are moral but also because—in virtue of being moral—they can be effective.

In the technological domain there is already a working precedent in the movement among people with disabilities for barrier-free design. The movement's achievements are now apparent in the profusion of ramps and modified restrooms in public places, and began becoming even more apparent with promulgation of new design regulations under the Americans with Disabilities Act of 1990. The movement has an oppositional component but also a clear moral basis and a constructive hopeful thrust. Nonmarket, democratic design criteria—often formulated and applied by disabled laypeople—are now being used to define individual and collective needs, including access to public space. Moreover, participants do not evaluate just one technology at a time—the

norm in conventional technology assessment—but entire technological and architectural environments.

Sometimes people with disabilities have described themselves as
the latest disadvantaged U.S. group, after African Americans and
women, mobilized to assert their human rights. That self-charac-
terization is fair and has proven effective. But it may also be selling
themselves short. In terms of their consciousness of technologies' nonfocal structural significance, and their impressive success in articulating
their concerns and in organizing to democratize technological orders,
haven't physically disabled citizens outdone every other minority and
the able-bodied majority too? Technologically speaking, it is the latter
who now seem most disenabled, but who fortunately have disabled
brethren as empowered role models.

Freedom of Research

One occasionally encounters the objection that participatory RD&D
would unjustifiably infringe traditional researchers', engineers', and
designers' freedom of inquiry.[58] This objection seems ill-considered on
several counts.

First, when couched as a matter of a constitutionally protected
freedom or right, it is easily overstated. There is no unlimited, nor clearly
demarcated, constitutional protection of research.[59] Second, research is
already socially guided and constrained in many respects. Researchers
in private firms and in government labs are much more often assigned
to projects or research areas than granted carte blanche to pursue
personal curiosities. Almost all researchers and designers are constrained by the social availability of funds to support RD&D in their
domain of interest. It is illegal to conduct certain forms of research (e.g.,
if it would callously harm human subjects or violate environmental
laws). U.S. medical research in novel, ethically uncertain areas is now
routinely subject to prior scrutiny or approval by an institutional review
board or ethics review board; university social science research is
similarly prescreened by human subjects review committees. Thus the
precedent for socially guiding research is well established.[60]

Third, and contrary to the objection's implication, participatory
RD&D promises generally to expand freedom of inquiry. By involving
more people, such processes will enlarge the social knowledge base on
which RD&D draws: "Lay involvement is not only good politics but
also good science."[61] By creating new avenues for disseminating research
results, it may create new opportunities for swift idea cross-fertilization.
As suggested earlier, working collaboratively with users and affected

nonusers promises to increase the satisfaction that researchers and designers feel in their accomplishments. It should also diminish managers' powers to suppress promising products that they fear might compete with established product lines or to suppress engineers' ideas for humanizing work processes. (One recent business school study reports that what American corporate engineers "tend to lack is . . . most important[ly], the power with which to make [their] visions real."[62]) Above all, insofar as freedom of inquiry is a good, participatory opportunities will vastly expand the pool of people positioned to exercise that freedom. Indeed, if the objectors' real concern is that they don't want to share their freedom with new social groups, their ethical case is shaky. Once upon a time, propertied white males were the only people entitled to vote, and they too objected to extending political participation on the grounds that it would be inefficient and infringe their own political freedom (i.e., their illegitimate social power).

Assuredly, participation could in some cases be misorchestrated in ways unnecessarily interfering with individual autonomy. But that is a risk against which to guard, not a sufficient reason to disallow participation across the board. Participation also invariably engenders some social frictions and frustrations, but their extent ought normally to be much less than the converse frustration, demoralization, and injustice associated with persistent denial of the right to participate. In any case, the examples of participatory RD&D reviewed here include no obvious instances of the oppression of participating professionals, many of whom learned to enjoy the process wholeheartedly.[63]

On the other hand, participatory RD&D could lead to swifter and more widespread popular understanding of emerging technological developments, in turn sometimes eliciting popular opposition. Isn't that precisely what one should hope? A democratic politics of technology must seek a broader and more democratic menu of choice, but then subject that menu to more discriminating democratic scrutiny prior to permitting extensive deployment and prior, if possible, to unnecessarily expensive developmental effort. If undemocratic technologies would thus run greater risk of early social detection, substantively democratic RD&D programs, which today often languish for lack of institutional support, would have a much greater chance of winning deserved public backing. A democratic politics of technology is not antitechnology, but prodemocratic technology.

Chapter 12

TECHNOLOGICAL POLITICS AS IF DEMOCRACY MATTERED

The precise form of future, strongly democratic sociotechnological orders, as well as the pathways to reach them, have to be worked out through trial and error and prolonged political effort. Like today's non-democratic processes, the arrangements for making democratic decisions about technology will necessarily be complex and will have to evolve organically. This is necessary to reflect the diversity of local circumstances, harness the creativity of democratic participation, and help secure the institutional resiliency needed for long-run societal sustainability.[1]

Because a detailed blueprint would be inappropriate, this chapter presents an illustrative set of tactics and strategic objectives. For the sake of concreteness, the focus is on the United States, looking forward from circumstances that existed in 1994. Of course, technological democratization is no less important elsewhere and in the more distant future, in which case alternative strategies may be appropriate. The recommendations that follow are listed roughly according to the increasing difficulty of implementing them on a strictly ad hoc, local basis.

AWARENESS AND MOBILIZATION

How might one start to broaden social awareness of the need to democratize technology and begin cultivating grassroots involvement?

1. Mapping Needs and Resources

One approach is to organize community projects to inventory local technologies and assess their social consequences, their combined compatibility with democracy, and opportunities for improvement. Apart from being practicable, an inventory would draw attention to aggregate nonfocal effects, countering the impulse to imagine that one need only introduce one or two focally democratic technologies into technological orders that otherwise remain substantially undemocratic. The movement among people with disabilities for barrier-free built environments offers a partial prototype for this kind of sociotechnological mapping. Mapping could provide a natural context for beginning to debate and apply design criteria for democratic technologies.[2]

Mappings can initially be directed toward an entire city or town or, much more modestly, toward a single neighborhood, urban block, or workplace. The initiators could be virtually any group—perhaps members of a senior citizens group, a church, a junior high school classroom, or a planning board. Extant projects to re-create local public space (e.g., a community garden or a new downtown pedestrian area) or to promote environmentally sustainable economic development could provide excellent action-oriented contexts for mapping.

2. Education, Research, and Publicity

a. Education

Another important, short-run step is to create new educational curricula for enhancing popular understanding of technological politics. Adult education classes on social control of technology at Britain's television-taught Open University represent one promising model; some of the workers who developed the Lucas Aerospace Alternative Plan had previously taken these courses.[3] In the United States, the nonprofit National Association for Science, Technology, and Society helps secondary school and college teachers develop instructional materials involving the social dimensions of science and technology.[4]

There are also a few farsighted programs to teach design skills to all students, not merely to those tracked toward one of the designing professions. Conversely, it is important to incorporate collaborative design efforts with laypeople into design professionals' curricula and to establish training and career opportunities for participatory design facilitators.[5] A freshman engineering design course under development at Rensselaer Polytechnic Institute takes a step in this direction by integrating diverse social concerns and perspectives on technology into all classroom discussion and design projects.[6]

Apart from requiring expansion, U.S. efforts in these areas can be strengthened. Initiatives to promote scientific and technological literacy remain perennially fashionable but, democratically speaking, they are seriously incomplete.[7] Some of the energy behind them could nonetheless be guided toward more democratic purposes and methods, as could present instruction in professional ethics, narrowly construed. Best of all would be explicit discussion of democratic politics of technology as a standard topic in secondary school and college.

Should education merely impart ideas or should it also teach active citizenship? Experiential learning can contribute to the latter objective. For instance, Worcester Polytechnic Institute (WPI), in Worcester, Massachusetts, requires all undergraduate students to undertake a project actively relating their technical education to a social problem. All of WPI's faculty members become involved as project supervisors.[8] Experiential learning also opens the possibility of using school projects as a vehicle to initiate broader community involvement in the democratic politics of technology.

b. Research

To what avail is launching new courses if it is not obvious what to teach? Graduate assistanceships and new programs of research oriented toward democratic technology are thus vital. Early on, this should especially include support for strengthened social scientific research—conducted, whenever possible, collaboratively by lay citizens and scholars—into the social history and anthropology of technology and into designing and evaluating projects for advancing technological democratization.[9]

Examples of topics meriting inquiry include: (i) how one type of technology can affect the use or adoption of others; (ii) conditions under which technological change transforms versus reinforces social relations; (iii) lessons from history concerning promising approaches to forecasting technologies' social consequences; (iv) social experiments to reveal the extent of latent flexibility in the social relations with which existing technological hardware can operate; and (v) successful and unsuccessful approaches to participatory decision making.[10]

The U.S. National Science Foundation sponsors pertinent scholarly research under its Ethics and Value Studies (EVS) program. To date, most EVS studies have involved research on medical ethics; environmental, health, and safety issues; or science and engineering education.[11] Projects attentive to technologies' broader or more subtle social, political, and cultural consequences are a distinct minority. This could begin to change if the National Science Foundation follows through on a proposed new funding initiative in science, technology, and democracy.[12]

Many EVS grants support research pursued within university science, technology, and society (STS) programs.[13] In recent years, STS faculty research and graduate training have tended to drift into theoretical preoccupations far removed from community and citizen needs or even from basic studies of technologies' social consequences. However, there are important exceptions, and critics in the STS field have called attention to the problem.[14] One part of the solution would be to press universities and professional societies to alter the criteria governing hiring, promotion, and accreditation to reward socially oriented research, participatory research, community activism, and popularization of ideas.

Some of the most democratically instructive STS inquiry has grown out of participatory research with technologically distressed communities. For instance, during the 1970s and 1980s, residents of Woburn, Massachusetts, became worried that toxic waste from nearby industries was causing leukemia in their children. Confronted by government skepticism and indifference, they were able to collaborate with scientists from Harvard University's School of Public Health in conducting their own epidemiological surveys. The Woburn case eventually influenced federal legislation on hazardous wastes and led to a multimillion-dollar legal settlement; moreover, sociological study of the episode produced pathbreaking insights into science, technology, and politics.[15]

c. Publicity

Competent, creative participation often depends on knowing about some practical sociotechnological and institutional alternatives. Alternatives can reveal the historical contingency of one's current situation and provide a kind of vocabulary on which to draw for formulating options. (Recall the U.S. automobile workers who began conceiving of ways to redesign their own workplaces only after learning of Volvo's innovative factory at Kalmar.[16]) Participation can also benefit from some acquaintance with social criticism of technology.

Media coverage of science and technology tends toward breathless accounts of amazing discoveries, punctuated by episodic outrage over the occasional technological calamity.[17] Major developments in national technology policy, when covered at all, most often appear as business stories focused on a technology's financial implications. In short, print and electronic media need encouragement to run columns, guest editorials, and documentaries reflecting social criticism of technology and dealing more thoughtfully and comprehensively with technological politics worldwide.

Whenever possible, it is desirable to escape the experts-enlighten-the-masses format. For example, the Canadian Berger Inquiry's community hearings, broadcast over radio and reported on television, were designed with the help of participating native tribes so that villagers would be able to testify in a nonintimidating setting. Televised citizen technology tribunals or community-produced video documentaries are other promising possibilities. These convey specific information but, just as importantly, teach viewers by example that people "just like me" can play a role in technological decisions.[18]

3. Political Movements

The idea of democratic technology is probably too diffuse and abstract to elicit a full-blown mass movement of its own. More plausible would be a smaller movement of dedicated groups and individuals sufficient to focus popular attention on the issue, to help forge larger ad hoc alliances, and to galvanize other organizations and movements into recognizing their interest in advancing technological democratization. The seeds of the requisite core group are evident in cross-issue national conferences concerned with science, technology, and social activism that were convened in recent years by the Institute for Policy Studies, the 21st Century Project, and the Foundation for Deep Ecology.[19]

A number of progressive social movements already have some experience with the democratic politics of technology. Examples include organizations working on environmental issues or environmental racism; coalitions supporting defense conversion or sustainable economic development; unions and workplace health, safety, and democracy activists; groups concerned with specific technological domains, such as telecommunications, biotechnology, energy conservation, and renewable energy sources; the disability rights movement; and so on. All of these would benefit from generally expanded opportunities to participate in technological decision making; already within some of these movements this interest is becoming explicit.[20] Cross-movement alliances are probably critical to mustering the political clout needed for firm progress. Within this context, efforts to increase democratic control over communications media and to provide fair access for underrepresented minority perspectives are especially important.[21]

There are other social movements that have not historically paid much attention to technology but whose core agendas are nonetheless affected by it. The women's movement, the civil rights movement, and urban renewal and housing groups are examples. Women's groups have, for instance, influenced medical research policy, but feminist scholarship criticizing a wide range of nonmedical technologies has generally not yet prompted political action.

How can organizations and movements decide whether technology matters enough to them to justify allocating a fraction of their scarce resources to technological politics? Students often enter university STS programs because they want to do something constructive about a social problem involving science or technology. A program of competitive internships to place STS students or graduates with appropriate public-interest or grassroots groups would address two problems at once. The students would learn about activism and social engagement by actually doing it. At the same time, they could work with their host group to explore ways that science and technology decisions bear on that group's existing agenda.

Sympathetic political leaders, government and corporate administrators, and engineers, scientists, and other professionals have helped nurture many examples of a democratic politics of technology. Their efforts have proven especially constructive when conceived as collaborating or facilitating rather than directing. Support for technological democratization may be strongest on the progressive side of the political spectrum. But democracy is a motherhood issue (few are willing to admit publicly they are against it) and strong democracy's respect for cultural tradition, democratic community, and political decentralization could appeal to many conservatives, too. But one can anticipate opposition among the elite circle of business, military, and academic leaders who have exercised hegemony over science and technology decision making since World War II.[22]

Can technological democratization possibly come about? One response is to consider the situation of ecologist Rachel Carson when she published *Silent Spring* in 1962. A skeptic might well have counselled Carson to give up. The ecological problems she portrayed were so pervasive, so entrenched in economic processes and power relations, and so beyond the everyday consciousness of most people that surely there was virtually no chance her voice would be effective. Yet within a decade, a major environmental movement was born, the first Earth Day took place, the Environmental Protection Agency was established, and the U.S. Congress enacted landmark legislation to begin improving air and water quality. Carson is widely credited as an initiator of the movement that produced such results and that carries on the struggle to this day; she helped precipitate a preexisting inchoate background of popular concern. Today's aspiring technological democrats are not in such a different situation.

On the other hand, existing technological orders are so extensively, expensively, and interdependently embodied in material artifice that it is inconceivable that they could be transformed comprehensively with great rapidity, no matter how great the political will. Present roads, automobiles, sewers, dams, power lines, telecommunication networks, building stocks, and so on cannot be wished away. But structurally significant technological transformation is entirely feasible, provided that it is pursued realistically. For instance, government policies can establish performance standards for new technologies (analogous, say, to automobile fuel efficiency standards). Whenever existing technological systems require repair, replacement, or expansion, concerted democratic effort can push for more democratic variants (e.g., consider environmentalists' successes in ensuring that new power plants built into existing electric grids meet more stringent pollution standards). One can also explore ways of using existing hardware in conjunction with more democratic organizational structures (as when factories have combatted the social drawbacks of assembly lines by reorganizing workers into partially self-managed teams). Finally, it is reasonable to contemplate rehabilitating or retiring selected technological systems when the democratic gain would be sufficiently large (this is analogous to the expense incurred in removing toxic asbestos from school buildings).

a. Foundation Support

It is crucial in these early stages that charitable foundations develop programs to support a democratic politics of technology, including some core nonprofit organizations dedicated to that end. Because the idea is relatively new, virtually no foundations fund directly in this area, although some have programs that support one aspect or another (such as democratic telecommunications policies). Foundations might also wish to reexamine their current programs to ensure that they are not inadvertently sponsoring antidemocratic technological developments.[23]

4. Time for Politics (and Life)

Can there be a self-governing society if citizens have limited ability to structure their own time? Programs of job sharing, flexible life-scheduling, benefits' provision to part-time employees, citizen sabbaticals, parental leaves, and affordable childcare and birth control need to be extended or created. The women's movement has promoted initiatives in these areas, and nationwide childcare is particularly far advanced in France and Scandinavia.[24] Other supporting steps would include expanding opportunities for part-time apprenticeships and for other forms of lifelong learning, while integrating more community projects and practical internships into the lives of youth; the Clinton administration has sought to advance programs of this kind.

In recent years many U.S. corporations have used productivity improvements as the occasion to shed large numbers of high-wage jobs. Meanwhile, the remaining U.S. workforce toils much longer hours than, say, its French and West German counterparts. This suggests a straightforward way to combat unemployment and yet free up time for leisure and politics: shorten the work week and share the remaining jobs.[25] The U.S. Department of Labor or state-level counterparts could be set the task of planning, with the assistance of citizen advisory boards and public hearings, the transition to more humane, flexible, egalitarian, and self-managed structures of time organization. Many of the necessary components are already in place in various nations and workplaces.[26]

Citizen sabbaticals are important for reasons noted in earlier chapters and also for their contribution to the knowledge repertoire discussed in Step 2c. Faculty sabbaticals; some U.S. companies' programs of granting employees leave to engage in social service; and competitive travel, research, and arts grants (such as Fullbright and Guggenheim Fellowships) are precedents that could be broadened to reach out to nonprofessional workers.[27] Although sabbaticals ought ideally to be available to people at all life stages, a good first step would be to expand the Peace Corps, VISTA, and the newly established National Service Program (which subsidizes college tuition in return for community service).

Employers are legally required to release a worker who is called to jury duty. Fair and effective citizen representation in political decision making will require legally establishing analogous work-release time for participating in other domains of civic responsibility (such as a number of the participatory mechanisms outlined in later sections).

On average, Americans spend one-quarter of their waking hours watching television, and the hours will likely increase if the proposed national information superhighway starts delivering more program choices. Could an insidious cycle be operating here? More people at home watching television means fewer people out somewhere else participating in—and thus sustaining—face-to-face social and civic activities. But each decline in social activity, in turn, diminishes peoples' incentive to turn off the tube (why go out if it will be hard to find something better to do?). Communities that have organized one-week moratoria on television watching have seen an immediate increase in social and civic life. Could one dare propose experimenting with voluntary television and video moratoria as one-evening-per-week communitywide events?

Establishing a more strongly democratic society requires more than transformations in the technological domain.[28] Probably most important would be greater economic equality—say, ceilings on personal wealth sufficient to prevent gross hyperempowerment and floors sufficient to prevent inhumane poverty or politically crippling insecurity. This is essential to strong democracy generally but also to more fully realize technological democracy specifically—for fairness in access to individualized technologies, market influence on research and development and on other investment decisions, and more equal stature in government and corporate policy decisions bearing on technology.[29]

Until the political will for economic equalization reemerges, struggles for technological democratization will be more difficult. The latter could, on the other hand, gradually contribute to evolving political conditions more amenable to reducing economic disparities. As Ullmann notes, "Equality is a possible *product* of political participation and not its precondition."[30]

CORE ACTIVITIES AND INSTITUTIONS

The next several steps involve institutionalizing the core activities of democratic technological politics: democratic technological design, choice, and governance. Figure 12-1 suggests how this chapter's recommendations might fit together. The diagram shows earlier microlevel steps building toward later, more ambitious ones. For instance, education, research, and publicity (Step 2) should broaden social readiness to participate in core activities (Steps 5–8). The dashed, returning arrow is intended to suggest that eventual macrolevel structural transformation—represented here partly as a product of prior grassroots effort—would, in turn, facilitate extension of the preceding core steps.

The order of these steps is not stamped in steel. Sometimes it will be easier to convert a local military contractor to civilian democratic production (nominally Step 10) than to move forward on citizen sabbaticals (nominally Step 4). Political opportunities open and close unexpectedly. That is not a problem.

However, is there a deeper dilemma? So long as a society's technologies are insufficiently democratic, how can there be hope of evolving strongly democratic political institutions? Existing technologies

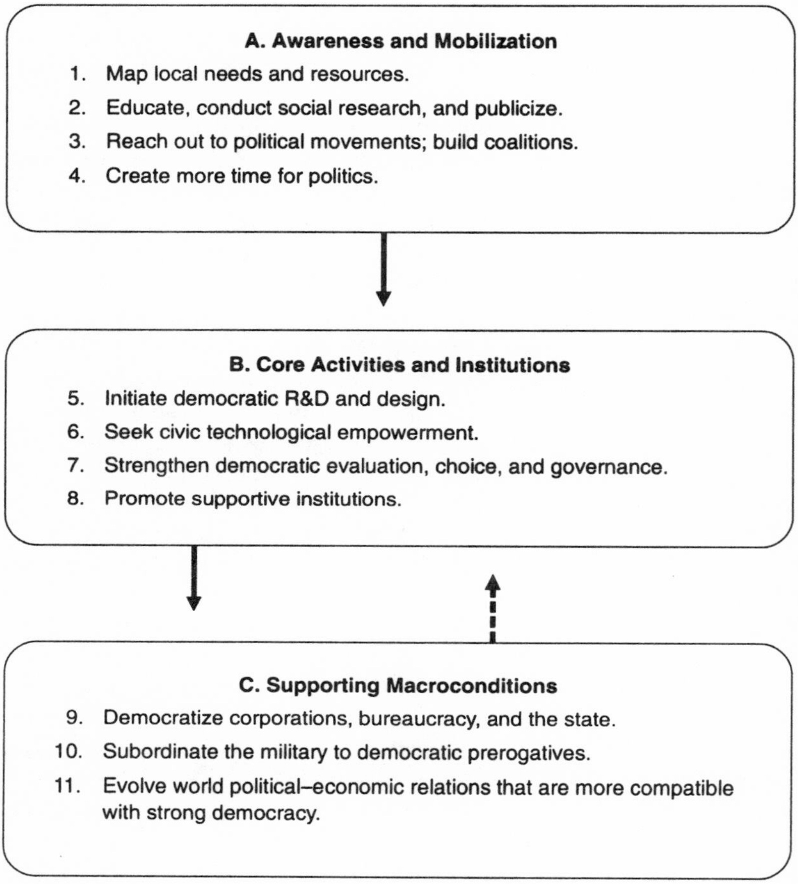

A. Awareness and Mobilization

1. Map local needs and resources.
2. Educate, conduct social research, and publicize.
3. Reach out to political movements; build coalitions.
4. Create more time for politics.

B. Core Activities and Institutions

5. Initiate democratic R&D and design.
6. Seek civic technological empowerment.
7. Strengthen democratic evaluation, choice, and governance.
8. Promote supportive institutions.

C. Supporting Macroconditions

9. Democratize corporations, bureaucracy, and the state.
10. Subordinate the military to democratic prerogatives.
11. Evolve world political–economic relations that are more compatible with strong democracy.

FIGURE 12-1. Institutionalizing a democratic politics of technology.

block the way to better institutions. But until more democratic institutions exist, how can there be hope of evolving democratic technologies? Current institutions block the way to better technologies.

The dilemma is not insurmountable. Figure 12-1 highlights political processes and institution building, but implicit throughout are accompanying technological transformations (prescribed provisionally in Part II). Present technologies and institutions are democratically imperfect but also allow space to maneuver and struggles for piecemeal improvement. (African Americans, women, and environmentalists have made real strides, despite a sociotechnological order often rigged against them.) In short, institutional and technological democratization can move forward untidily together.

Although some of the recommendations that follow are sweeping, most institutional innovations should, when feasible, be tested, evaluated, and modified via smaller pilot projects before being gradually adapted for widescale use. Participation itself must be structured with due attention to strong democratic norms. It ought to incorporate processes for collective self-education, real power to make or influence decisions, adequate accountability for actions, protection of minority rights, and fair representation of disadvantaged social groups.[31] When initial attempts fail, that should occasion learning and renewed effort to discover better approaches.

5. Democratic Design Processes

One core element of technological democratization is to vastly broaden chances to shape the agenda of choice. This implies extensive opportunities for citizen involvement in technological research, development, and design (RD&D) and strategic planning within universities, corporations, architectural firms, and government agencies and laboratories.

Diverse objectives each demand somewhat different institutional arrangements. There must be procedures allowing communities, groups, and citizens—including those today least empowered—to help directly initiate some RD&D programs and design technologies responsive to their needs. Democratically monitoring other ongoing technology development programs is important to permit early public involvement in emerging trends. Means must also be explored for countering the tendency of experts to cow participating laypeople. Finally, in order not to stifle local initiative, creativity, and serendipitous discovery, it is crucial to avoid an overly heavy-handed, centralized approach to managing societywide RD&D.

a. Democratically Conducted RD&D

As a start, one might encourage individual universities, businesses, or industrial consortia to undertake selective experiments in participatory RD&D. From a corporate perspective, this might initially be interpreted as an extension of market research; but unlike consumer surveys, direct participation would provide an avenue for considering the aggregate social effects of individual purchases. Universities could explore involving laypeople in ongoing faculty research.[32] Existing, nonprofit architectural Community Design Centers would benefit from government support and further replication.[33] Step 1's community mapping projects could provide one source of project ideas and design criteria.

As a follow-on step, government-supported research and development (R&D) could begin requiring grantees and national laboratories to include experiments in broadly representative, paid lay participation. Current R&D tax credits could be expanded and increased for corporations that involve workers and communities in R&D and strategic planning, make an effort to use democratic design criteria in their R&D programs, or undertake R&D in democratically prioritized areas (such as environmental sustainability). If desired, any resulting loss of tax revenue could be made up by lowering the R&D credits available to other corporations.

To increase social influence on the direction of research, some R&D incentives could be awarded through socially judged proposal competitions. For instance, currently the Advanced Technology Program, administered by the National Institute of Standards and Technology, funds private-sector industrial research based on a two-stage decision process. Technical experts are asked to rank competing proposals based on technical merit. Afterward, outside business experts judge the proposals' business and economic promise. No great conceptual leap is required to imagine expanding the latter review to include social and environmental considerations and inviting workers, public-interest group representatives, and state and local government officials to serve as evaluators.

In the 1980s, state governments across the United States launched programs to promote innovation-driven economic growth.[34] All such programs could be redesigned to become responsive to technologies' nonfocal social dimensions and to encourage experiments in participatory RD&D. There is a prototype in Massachusetts's quasi-public Center for Applied Technology (CAT). During the late 1980s, CAT launched a program to help firms develop or adopt new production technology. To benefit, firms had to pay workers to participate in CAT's consultation process and agree that any innovations would seek to enhance worker skills and reduce corporate hierarchy. For instance, CAT helped one 400-employee maker of packaging equipment alter its manufacturing process. Workers and their union played a major role in the design process, agreeing together with management to fulfill several basic criteria (e.g., skills had to be enhanced and parts throughput time had to be reduced). As a result, production time for spare parts dropped from several weeks to a few days, while jobs were redesigned to broaden workers' skills and increase their responsibility.[35]

This is a concrete instance of a government-assisted, participatory design process incorporating negotiated democratic design criteria. It could serve as one model for reorganizing the newly emerging, national network of manufacturing extension centers.[36]

b. Democratically Monitored RD&D

Are there precedents for social scrutiny of planned and ongoing RD&D efforts? Yes; examples include medical ethics review boards, municipal planning commissions, and local design review boards. Expanded to other domains, restructured so that their membership and performance better match strong democratic norms, and making explicit use of democratic design criteria, these or related mechanisms can provide a kind of hands-off oversight.[37]

Examples of worker participation in corporate RD&D are growing, and the Clinton administration is committed, in principle, to extending the practice. But, to date, the administration's actions remain contradictory. The Labor Department is charged with encouraging worker participation in developing and adopting new technology. Meanwhile, actual manufacturing technology development and transfer programs, administered by the Departments of Defense and Commerce, neither mandate nor even encourage worker participation in the projects they fund. (The Commerce Department has at least considered encouraging labor–management cooperation in developing new production technologies under its Advanced Technology Program, but so far no actions have been taken to implement this idea.[38])

It would also be possible to require direct, hands-on community involvement in all RD&D teams. But some caution is in order. The most inventive businesses are often small; thus mandatory public oversight could hamper their creativity, putting them at a competitive disadvantage. Moreover, large corporations currently command most of the resources needed to translate basic inventions into marketable products. A possible compromise would require citizen oversight or participation in large corporations' technology development efforts, but exempt smaller corporations. It would also be feasible to incorporate citizen oversight into the background institutions supporting networks of small manufacturers.[39] Early warning, even if limited to major developmental efforts, would be worthwhile if it could help reduce significant democratic misinvestments and unnecessarily bitter and costly downstream political battles.

Mechanisms are needed to ensure that oversight reflects the range of people and communities that an innovation would affect. For instance, worker representatives in affected industries should help oversee the development of new production technologies. Meanwhile, democratically adverse technologies emerging from smaller corporations or from other research centers could still be debated prior to extensive deployment (see Step 7a). Small corporations that wished voluntarily to be democratically monitored—perhaps to help limit political uncertainty—could request it.

c. Trade Secrecy

Wouldn't oversight of corporate RD&D be seriously hindered by trade secrecy law? A principal intent of such law is to encourage innovation by allowing businesses to monopolize commercialization of their own discoveries. A disadvantage, apart from restricting democratic involvement, is that it allows wasteful duplication of effort—corporations may not know what each other are doing—and hinders cross-fertilization of ideas.[40] Trade secrecy can, moreover, be regarded as a significant infringement of employees' freedom of expression—particularly once one understands that technological secrecy may conceal information vital to democratic decision making concerning a society's basic structure. It can also be morally or socially corrosive to compel people to keep trade secrets. High divorce rates in California's Silicon Valley have been attributed partly to security barriers that prevent spouses from knowing what their partners do at work.[41]

On balance, it seems desirable to relax trade secrecy and, as needed, provide compensatory incentives for ongoing investment in innovation. Many approaches are conceivable.[42] The democratic justification for revising intellectual property law is strong; the present laws are predicated on the assumption that virtually all technological innovation is desirable, and thus give no consideration to innovation's nonfocal, structural social consequences. The need for democratic reform is growing, because trade secrecy and patent protection are increasingly being extended to both academic and government research programs.[43] Strategically, one must anticipate that many corporations accustomed to operating under the current legal regime will oppose democratic reforms.

d. RD&D Policies

The preceding recommendations focus on decision making in individual research institutions. However, broad R&D trends are also established by setting priorities at the state and national level through intricate, but only loosely coordinated, political and bureaucratic processes. Should government R&D policy be developed through an openly democratic process that would encourage democratic technology and participatory RD&D?

Why pursue more extensive public participation in decisions that we already elect legislators to make? The central problem is that few legislators have time to become familiar with the details of science and technology policy. As a result, they routinely delegate countless important decisions to nonelected staff, administrative agencies, and advisory boards. Staff, agencies, and boards, in turn, depend extensively on technical experts.

The rationale for privileged political involvement is that accomplished scientists, engineers, economists, and business experts are often the best judges of technical feasibility and promise. The shortcoming of the rationale is that experts are neither impartial nor particularly well qualified when it comes to assessing the social and political significance of proposed research, and yet judgments of the latter sort are inextricable from judgments of technical feasibility.[44]

Stronger citizen participation in setting R&D priorities involves several difficulties. One is that lay citizens often are not well qualified to judge research on its technical merits, and thus experts must continue to play a role. Second, when an advisory board includes both experts and laypeople, lay members may be marginalized or intimidated, especially if they are in the minority.[45] Why not explore innovative, hybrid advisory mechanisms, such as (i) a system of dual expert and lay advisory panels or (ii) lay panels, supported by participation facilitators, that take testimony from competing experts with expert cross-examination? The U.S. National Institutes of Health already include nonscientists in establishing priorities among competing R&D funding domains.[46]

Sweden's high-level Council for Planning and Coordination of Research, which funds emerging research areas, goes further. A majority of the council's members are laypeople. Besides a half-dozen leading scientists, there are five members of Parliament, representatives from Sweden's four national union combines (three for labor, one for employers), and three at-large members (such as journalists and local government officials). The council was established in this form precisely in order to permit public influence over national research priorities. It is noted for innovative approaches, such as supporting research on alcoholism that integrates medical and social science perspectives.[47]

In 1992 an independent Commission on Science, Technology, anu Government established by the nonprofit Carnegie Corporation proposed creating a standing, nongovernmental National Forum on Science and Technology Goals in the United States. Working in the context of the challenge to long-standing national science and technology policies introduced by the end of the Cold war, the commission envisioned that

> the Forum would convene individuals from industry, academia, nongovernmental organizations, and the interested public to explore and seek consensus on long-term [science and technology] goals and the potential contribution of scientific and engineering advances to societal goals.[48]

In 1994 the National Science Foundation awarded a grant to support exploratory work on a decentralized democratic variant of this proposal.[49] A decentralized forum would be an excellent place to begin involving a broad range of people in science and technology policymaking, including the debate of competing design criteria for democratic technologies.

It is far from obvious that technological democratization requires any crash programs of basic natural science or engineering research. Certainly, such research may yield important insights. But there are countless ways to advance technological democracy through development, design, and organizational research applying the existing stock of basic scientific know-how.

Step 5 envisions broadening the menu of democratic technological alternatives. The next two steps would subject proposed technological deployments to more careful democratic scrutiny. This should establish a climate of political and economic expectations that would feed back to influence microlevel R&D decisions. Thus Steps 5, 6, and 7 would, eventually, function interdependently.

6. Civic Technological Empowerment

Following through on earlier steps and experiences, communities can begin responding to their identified needs for a more democratic technological order. When required, there should be state or national assistance in helping localities democratically plan, finance, and implement:

i. Sponsoring or undertaking of necessary social science or hardware-oriented RD&D;

ii. Reestablishment of local democratic governance of technological infrastructures;

iii. Discouragement or replacement of democratically adverse technologies;

iv. Re-creation of public spaces and deployment of civic tech-
    nologies and architecture supporting democratic community
    and local and translocal democratic politics; and
v. Monitoring of local economic dynamics in order to be able to
   intervene constructively when democratically adverse structural trends develop.

One need not envision democratic planning as an alternative to
market interactions but as a means of ensuring that market performance
remains compatible with democratic goals and structural requirements.
For example, democratically revised zoning ordinances, building codes,
environmental regulations, tax and financial incentives, or consumer
warning labels can help guide private decisions toward desired social
outcomes without dictating specifics of how people should lead their lives.

Apart from formal local government institutions, other institutions
that could—if made democratically accountable and representative of
the entirety of a community—facilitate various aspects of civic technological empowerment include community development corporations,
community land trusts, housing associations, municipal utility companies, the secondary institutions supporting networks of small manufacturers or cooperative workplaces, and so on.[50]

Both the means and the results of technological empowerment will
presumably vary greatly among communities, which is desirable from
the standpoint of societywide cultural pluralism and freedom of choice.
In daily life people would continue to seek the jobs they want, make the
purchases they want, and arrange their free time as they want. The great
difference would be a reasonable assurance that institutions were operating to ensure that the collective consequences of diverse personal
decisions remain consonant with perpetuating democracy and other
shared values. Citizens would therefore know that when they pleased
(or when sufficiently displeased) they would have ample opportunity to
participate directly in those guiding institutions.

Pushed to excess or misconstrued, could civic technological empowerment ever infringe too deeply the prerogatives of daily life and
personal choice in particular communities? That is a danger against
which to remain vigilant, a principal reason for seeking nonintrusive
policy instruments, and potential grounds for establishing certain
legally protected zones of personal action. But this danger must be
balanced against the converse certainty that some of the personal
freedom people experience today is much less than it seems, in the
sense that latently it both reflects and perpetuates considerable structural inimicality to democratic empowerment, sociability, fairness,
and individual self-actualization.

There have recently been two basic strategies for implementing state and local civic planning in the United States. Most common are corporatist approaches, in which government leaders and powerful private interests formulate plans and broker deals among themselves. Under this system, there is minimal democratic involvement or openness, and the results are often inequitable.[51] In pragmatic terms, sometimes (e.g., when grassroots progressivism is not well developed) elite-brokered planning can be better than none at all. However, federal or state governments could help by at least requiring open brokering, public hearings, and accountability in return for local block grants.

Democratically more promising are progressive grassroots movements and political coalitions—like those in Burlington, Vermont, and Santa Monica, California—that have used local electoral victory as a platform for developing institutions for open participatory planning and civic revitalization. Santa Monica subsequently introduced an experiment in teledemocracy: citizen access to government information, services, and decision-making processes via widely distributed computer terminals. Encouraging steps toward civic technological empowerment are also evident in the emerging movements and coalitions supporting defense conversion, environmental justice, and more environmentally sustainable regional economies.[52] For example, responding to grassroots pressure, the state of Washington established a citizen advisory group to monitor military spending in the state, assess post–Cold war economic needs and opportunities, and promulgate action plans to help defense-dependent communities diversify their productive base. The board includes representatives from state agencies, local governments, the military, the defense industry, peace groups, and labor unions.[53]

One method to encourage ongoing democratic planning would be to sponsor periodic civic design competitions, with public funding to support participation by disadvantaged constituencies and neighborhoods. After public discussion of rival plans—including use of democratic design criteria—an initial nonbinding referendum would allow voters to rank-order the various design alternatives. Afterward, competing teams could revise their original plans in preparation for a final vote.[54]

Some aspects of civic technological empowerment may require higher level legislation enhancing localities' legal powers.[55] However, even without overall democratic civic empowerment, significant measures can be taken to help democratize local technological orders. For instance, landscape architect Karl Linn has helped many impoverished urban neighborhoods combine volunteer labor with donated or discarded materials to transform abandoned lots into playgrounds, parks, and public gathering spaces.[56] There are also examples across the United States of local citizens and workers negotiating with corporations the right to oversee safe management and reduced use of toxic chemicals.[57]

a. Economic Self-Reliance

One key to civic empowerment is to promote institutions and technologies favorable to more local production for local use. Apart from its other social and environmental benefits, this can be strategically vital because communities that become more economically diverse and self-reliant achieve some insulation from the coercive threat of corporate capital flight.

Efforts in this direction have been pioneered by organizations such as the nonprofit Institute for Local Self-Reliance (ILSR). In the 1980s ILSR helped St. Paul, Minnesota, implement a strategic plan for self-reliance that included prominently labeling locally owned businesses and local products, establishing a fund to help finance new local businesses and give the community a stake in them, and assisting the commercialization of locally invented waste recycling technologies (e.g., an industrial process for turning worn tires into new rubber or plastic products). A desire to keep local dollars from leaking out of the local economy, and thus capture economic multiplier benefits, led to an ambitious program of energy efficiency improvements. The program included creating a district hot water and heating system that by 1990 was the largest in the United States, providing heat to more than half the buildings in the city's central business district.[58] Beyond advancing self-reliance, the latter initiative exemplifies development of a more locally governable infrastructural system.

There are many other strategies for promoting a more self-reliant economy. For example, in the northwestern United States a publicly supported, nonprofit organization named Oregon Marketplace uses an import-substitution strategy. Oregon Marketplace has implemented a computer-assisted system that matches Oregon firms that import production inputs from out of state with competitive bids from potential in-state suppliers. As of 1992 each dollar that the state invested in this program was generating an average of 40 dollars in new contracts for Oregon businesses—in conventional economic terms, more than paying for itself in expanded tax revenue.[59]

Farming is big business in the state of Indiana, where 80 percent of agricultural sales come from two crops: corn and soybeans. Yet because of the lack of crop diversity, this farmbelt state actually imports most of the food that its citizens eat. Faculty at Ball State University, in Muncie, Indiana, have begun regional market studies, satellite-aided ecological analysis, and promotion of local farmers' markets to help

Indiana farmers diversify their crops, reduce chemical inputs, and sell locally. During its first phase the project identified 12 imported crops that could be grown locally, including asparagus, mushrooms, and (for zoos) bamboo.[60]

With proper legislative authority, it might also be possible to use flexible local or state tariffs to help balance the goal of greater self-reliance against the economic benefits of translocal competition and trade. Tariff revenues could then help finance other aspects of civic technological democratization.[61] A somewhat more visionary but less bureaucratic means of encouraging local investment and consumption is to deploy local currencies. Since 1991, residents of Ithaca, New York, have been able to earn and spend yellow and pink dollar-sized notes dubbed "Ithaca Hours." Paul Glover, the system's founder, argues that

> there isn't enough money in the local economy to do what people want to do. . . . With Ithaca Hours people can get an opportunity to earn more spending power, and the money doesn't leave Ithaca to buy rain forest lumber.[62]

Communities and states also need to seek democratic mechanisms for guiding local investment more toward the local economy, including into democratic technology initiatives and community- or worker-owned enterprises.[63]

In many respects, democratic civic technological empowerment lies at the heart of technological democratization. Thus the following steps, while often politically more ambitious, are intended to complement local self-governance, not to supplant it.

7. Democratic Evaluation, Choice, and Governance

Some remaining institutional gaps include:

  i. Establishing democratic regional, state, and federal mechanisms for helping to evaluate and democratize translocal technological orders;
 ii. Monitoring and guiding synergisms among complexes of focally unrelated technologies; and
iii. Overseeing dynamic structural consequences of translocal technological innovation and market performance.[64]

This should be done with openness and participatory opportunities and also without overstepping local autonomy. Translocal intervention is warranted primarily when local decisions produce translocal structural effects, when minority rights are violated, or to help coordinate interlocal relations.

One basic need is a nested system of democratic procedures for debating and using democratic design criteria. The nesting should allow many individual and local choices to be decided autonomously, but enable a shift to progressively more inclusive and careful scrutiny when important democratic values or structural considerations are at issue. Eventually, every polity and group ought to have easy access to one or more public forums—with outreach to disadvantaged groups—where citizens know they can go to deliberate effectively concerning technologies' social dimensions. Possible mechanisms include at the base a system of local or regional public citizen tribunals (see below), commissions, or hearings, which then feed into more formal state and federal government proceedings.[65] For instance, taking into account advice from a decentralized National Forum (Step 5d), the Congress might establish a basic legal and procedural framework for formulating democratic design criteria and then ask state and local governments to develop democratic mechanisms for extending and adapting general criteria to local circumstances.[66]

There is no single best model of expert–layperson relations within decision-making forums. One option mentioned earlier is participatory research, in which laypeople and experts collaborate on roughly equal terms.[67] Alternatively, contending interested parties—including both lay and expert representatives—can be asked to strive for negotiated consensus or, failing at that, to produce a reasoned dissensus. The democratic virtue of reasoned dissensus is that it helps others understand the bases of evaluative disagreements, rather than sanctioning behind-closed-door compromises that obscure those bases.[68] The concurring and dissenting opinions written by judges on the Supreme Court offer a familiar example of this practice.

In yet another model—citizen tribunals—groups of laypeople reach their own judgments based partly on testimony from competing experts.[69] In 1992, for example, a panel of ordinary Danish citizens attended two background briefings and then spent several days hearing diverse expert presentations on genetic manipulation in animal breeding, in a forum open to the public and journalists. After cross-examining the experts and deliberating among themselves, the lay panel reported to a national press conference their judgment that it would be "entirely unacceptable" to genetically engineer new pets but ethical to use such methods to develop a treatment for human cancer.[70] Their conclusions influenced subsequent parliamentary legislation.

To organize this type of "consensus conference," the Danish government's Board of Technology selects panels of citizens of varying backgrounds and then publicizes their judgments through public media, local debates, leaflets, and videos. In the case of biotechnology, the board has subsidized more than 600 local debate meetings. Dr. Jørn Ravn, the board's general secretary, explains that "citizens alone are the final judges of what they find good and promising, insufficiently examined, or perhaps even totally unacceptable." Research suggests that the Danish public and politicians are better informed on issues addressed this way than are the citizens of other countries facing similar questions.[71] Its achievements have caused the Danish process to be emulated recently in the Netherlands and the United Kingdom, and further emulation is under consideration in other European nations and under the auspices of the European Union.

The Danish process is a specific implementation of a general model in which (i) technical experts, (ii) experts in technologies' social dimensions and effects, and (iii) representatives of organized interest groups (including public-interest groups) play vital roles, but final judgment is left in the hands of representative everyday citizens. (Experts in social consequences are important for their ability to stimulate others' imaginations and to raise critical issues otherwise apt to be neglected. However, the obvious value-ladenness of the issues such experts address—plus their own expert blinders and preoccupations—render them unsuitable proxies for the citizenry as a whole.)

In contrast, in the United States a more common mode of technological judgment is one in which the great majority of participants are technical experts or representatives of organized stakeholder groups; thus experts in technologies' social effects and everyday citizens are outweighed or, more often, excluded entirely. A central limitation of this model is that the aggregation of technical expert and stakeholder views is apt to greatly slight technologies' broader social and political consequences. For instance, when—as is often the case—the represented stakeholders include business people, workers, and environmentalists, then economic, workplace, and ecological concerns will normally be addressed. That is good. However, nobody is there to watch out for cultural repercussions, structural political ramifications, or the overall public good. On the latter issues, variants of the Danish model appear much more promising.

In adapting such a model to the United States,[72] one might worry that consensus is much easier to achieve in a small homogeneous nation such as Denmark. That is true, but in terms of strong democratic objectives the important feature of the model is its efficiency in cultivating informed citizen judgment, even if the final report represents a reasoned dissensus.[73] (Besides, consensus will not invariably prove impossible; U.S. juries routinely reach consensus within the context of highly contested, complex legal disputes.)

It is also true that a single lay panel composed of, say, 15 people is a feeble statistical sample of the entire United States. However, the assembled groups are not being asked to promulgate binding laws or regulations; their deliberations are merely advisory to the public as a whole and to elected officials. In that context, hearing the considered views of a diverse group of 15 everyday citizens would be a marked improvement over hearing from none (which is the norm in a great deal of contemporary technology policy analysis and decision making). Moreover, on especially important issues, one could seek greater representativeness by assembling a succession of small lay panels or a single, larger group.[74] In any case, given prevailing U.S. disparities in wealth and overbusy lives, both fairness and efficiency would mandate paying people to participate.

a. Impact Statements and Social Trials

Federal actions affecting the environment must be preceded by an Environmental Impact Statement (EIS). Prior to a risky medical procedure, doctors must obtain a patient's informed consent. Yet U.S. law ironically permits the introduction of many technologies having the potential for profound societywide impacts without any evaluation of their social or political ramifications.

Public-interest groups, Congress, or the president could begin redressing this omission by asking federal agencies to comply with an existing but little heeded federal regulation. Promulgated by President Carter's Council on Environmental Quality, the regulation mandates attention, within current environmental impact assessment procedures, to a proposed technological installation's "aesthetic, historic, cultural, economic, social, or health [effects], whether direct, indirect, or cumulative."[75]

Even when rigorously enforced, however, current EIS regulations only apply to "major federal actions"—a triggering threshold not always crossed in conjunction with socially significant, private-sector technology innovations. Thus, to fully institute informed consent at the societal level, there should be a law requiring corporations and government agencies to publicly file a succinct Social and Political Impact Statement (SPIS) prior to introducing or importing a significant technological innovation. A notable new technology or design innovation (such as a biotechnology breakthrough) ought to be accompanied by a generic SPIS delineating potential and probable impacts, within all reasonably foreseeable contexts of production and use. In addition, construction of major technological or architectural installations (e.g., large power plants, shopping malls, or telecommunications systems) ought to be preceded by a context-specific SPIS. To help ensure responsiveness to societal concerns, there could be a requirement to empanel representative groups of potentially affected citizens to help prepare or oversee production of each SPIS.[76] (Here it is vital to remember that the range of "affected citizens" almost invariably extends far beyond those who are producers or direct users.[77])

An SPIS should highlight focal and nonfocal structural social consequences, including broader impacts that might occur if a technology and its supporting infrastructure become widely deployed. Legislatures could devise democratic procedures for specifying a minimum set of democratic and other social design criteria that each SPIS must consider. When social consensus on criteria proves elusive, SPISs could be required to address competing sets of partly contradictory criteria.

Various measures can help prevent counterproductive bureaucratization or unwieldy costs. For instance:

i. States might be allowed to experiment with alternative triggering criteria for identifying an innovation as significant enough to warrant an SPIS. Firms should be able to appeal for exemption if, for example, they can argue persuasively that an innovation is socially inconsequential or that its impacts are essentially no different from those of other innovations for which an SPIS has already been filed.[78]

ii. Business groups or consortia could be permitted to file a joint SPIS governing a range of innovations that are expected to produce similar or interdependent social consequences.

iii. Nonexempted small innovators or importers could be eligible for public grants to help defray SPIS preparation costs.

iv. It might be practicable to exempt a firm from the SPIS requirement if it has made a good-faith effort to permit early societal warning during its prior RD&D processes (Step 5b).

v. Finally, if SPIS preparation seemed apt to impose arbitrarily uneven costs on competing firms or to unacceptably impair domestic exports, the decision could be made to finance SPIS's out of general corporate tax revenues.

If the knowledge base for anticipating an innovation's consequences is weak, voluntary social trials in diverse communities could generate some of the missing evidence.[79] Especially in instances when a firm would in any case want to undertake market trials prior to full-scale production or deployment, democratic evaluation of the accompanying social effects would normally represent only a modest add-on cost.

The specific function of SPIS's could range from merely informational to providing a direct basis for further political oversight or intervention. For instance, an SPIS prepared during the early stages of development and design could provide a basis for socially altering a technology's developmental trajectory.[80] On the other hand, in many cases—and perhaps especially when an impact statement is prepared after a technological design is far along toward completion—any political decisions regarding a technology's use ought to be made and implemented locally.[81] This would fulfill strong democratic procedural norms and also be consistent with the context-specificity of many technological consequences.

Political intervention need not always be restrictive. For instance, innovations promising unusual democratic or social benefits, but for which the initial commercial market potential is slight, might warrant financial or regulatory assistance or targeted government procurement. There is already a precedent in federal "orphan drug" legislation, intended to encourage commercialization of promising medications for rare illnesses.[82]

But one must learn to evaluate warily optimistic assertions that a technology will much enhance democracy. Such claims often mistake a mere possibility for a contextual probability, while overlooking aggregate latent effects that can swamp in social significance a technology's focal purpose.

b. Integrative Technology Assessment

Technology-specific SPIS's need to be complemented by social studies of (i) the synergistic interactions among focally unrelated technologies, (ii) entire technological orders, and (iii) the combined consequences of diverse government technology programs. One means would be to require the U.S. Congress's Office of Technology Assessment (OTA)—or other appropriate federal or state agencies—to conduct some of the requisite research. The OTA might begin by using just one of Part II's proposed democratic design criteria at a time, but applying it comprehensively across an entire regional technological order. When sufficiently refined, integrative assessments of diverse technological orders could provide a guiding conceptual context in which to situate SPIS production for individual technologies (Step 7a). Conversely, SPIS's and other sociotechnological studies prescribed in earlier steps would provide information vital to generating more comprehensive, synthetic studies.*

Second, all OTA studies should include technologies' compatibility with democracy as a highest order evaluative consideration.[83] Third, why not revive earlier OTA experiments with lay participation in technology assessment,[84] including experimenting with variants of the Danish consensus conference process and the Canadian Berger Inquiry? The latter processes allow laypeople much more influence and public voice than do current OTA report advisory processes, greatly increasing the incentive to participate. Such explorations should include some projects focused on the technological needs of disadvantaged social groups and communities.[85]

Fourth, the OTA normally strives to produce consensus reports calculated not to test the limits of congressional opinion. While politically prudent, this can result in blandness or superficiality. "OTA policy analysis," says one former OTA project director, "is often too 'safe.' "[86] Others observe that controversial reports are more likely to attract public notice and thus be used.[87] Thus, as an alternative, the Congress could ask OTA to experiment with sponsoring competing teams of value-partial assessors to publish independent studies on the same subject.[88] Finally, the OTA or a suitable national laboratory could be asked to compile summaries of the various studies and impact statements recommended in previous steps and to report publicly on recurring threats and opportunities bearing on technological democratization.

c. Government Technology Policymaking

Today direct "public" participation in establishing national science and technology policy does occur, but it is limited almost entirely to elite representatives of three constituencies: the Defense Department and the national weapons laboratories, the academic scientific research community, and business. Outside experts participate, for example, through the National Research Council; by sitting on a wide array of government advisory boards; and as government consultants, witnesses at congressional hearings, and peer reviewers in the award of competitive research grants. One observer likens the resulting narrow range of opinion to "the sound of one wagon circling."[89]

The poor representation of other societal points of view is exemplified in hearings, organized in 1992 and 1993, by the House of Representatives' Committee on Science, Space, and Technology on a compre-

---

*Late in 1994 some congressional Republicans proposed abolishing the OTA. If this is done, alternative institutional settings would have to be sought for the functions suggested in this section.

hensive proposed National Competitiveness Act. Of 120 speakers invited to testify during the course of 30 hearings, only two witnesses represented public-interest organizations.[90] (Both were analysts from conservative think tanks, invited by Republican-minority congressmen who opposed the act as an example of unwarranted market intervention.) Thus, there was not even one witness from an environmental, defense-conversion, or labor organization on a major piece of legislation with extensive environmental, employment, and other social implications.

Throughout the 1980s there were repeated calls for rationalizing federal management of science and technology policy by establishing a cabinet-level department or a civilian counterpart to the Pentagon's Advanced Research Projects Agency. The Clinton administration moved in this direction by creating an interagency National Science and Technology Council—which the president himself chairs—and by allocating new responsibilities for civilian technology development to the Commerce Department. The potential effectiveness of such a strategy has been debated for years, but it is incontestable that most versions—including the one now being pursued—are democratically insensitive.[91]

An alternative approach would channel reform impulses into democratizing or decentralizing many aspects of federal science and technology decision making. All government or government-financed science and technology advisory panels, pertinent congressional hearings, and programs to develop or disseminate new technology need robust representation of many viewpoints, including those of everyday workers, consumers, and citizens. For instance, federal agency advisory boards—such as the Defense Science Board—ought to include representatives of affected public constituencies and be obligated to solicit additional advice through a decentralized National Forum (Step 5d), Danish-style tribunals, or other effective participatory mechanisms.[92] Selected aspects of federal policymaking authority can also be delegated to democratically structured regional or local authorities. This was done early on in regulating recombinant-DNA research.[93]

Here the lessons of any failed prior efforts at orchestrating participation need to be heeded. For instance, in the late 1970s the Carter administration solicited wide public comment on its proposed National Energy Plan. In danger of suffocating beneath an ensuing avalanche of 27,898 written responses, the administration shoveled them away into boxes, largely unread.[94] Similarly, there are few more effective ways to spawn cynicism than by inviting citizens to attend public hearings that do not demonstrably incorporate their concerns into subsequent decisions.

Even well-intended proposals to target national research more directly toward social needs often fail to incorporate an adequate grasp of the aggregate nonfocal means through which technological complexes exert social influence.[95] To cite an obvious example: by themselves "green technologies" can do little to promote a more environmentally sustainable economy if their benign effect is outweighed by the ecological harm and resource depletion caused by other technologies. Moreover, if not designed and deployed with a wider set of social values in mind, green technologies can also contribute nonfocally to frustrating other, nonenvironmental objectives. (Recall how, historically, some successful water and sewage management systems impaired local self-governance.[96])

Socially conditioned investment incentives—as proposed earlier in the case of R&D tax credits (Step 5a)—could be one way to elicit and implement need-specific technology programs that are comprehensive enough to accomplish their focal aims while also remaining attentive to other important social objectives. The Greater London Enterprise Board adopted such methods in subsidizing small business startups during the 1980s.[97] This approach has the potential to be extended to any government program intended to promote or guide technological development, adoption, or use. For example, investment credits for business equipment purchases could be crafted to reflect various social and environmental concerns in addition to familiar technical and economic considerations.

All government policies involving technology need to be reevaluated from the standpoint of their implications for achieving a more democratic technological order. National technology and trade policies, for instance, are mostly predicated on helping U.S.-based firms compete in global markets. These policies need to be reviewed to make them more compatible with the complementary objective of achieving greater local self-reliance. There is particular irony in administration and congressional enthusiasm for promoting the international export of American-made environmental technologies.[98] These programs not only run counter to self-reliance both at home and abroad but also overlook the pollution and resource depletion associated with long-distance shipping. Certainly, it is preferable to export clean technologies than dirty ones, but from the standpoint of strong democratic norms it might make even more sense to encourage licensed production in the regions where clean technologies will actually be used or to assist other nations' efforts to develop, manufacture, and use their own environmental technologies.

Federal agencies such as the Nuclear Regulatory Commission and the Departments of Defense and Energy have begun to establish local boards to advise environmental cleanup projects. That is one promising approach toward coping with the legacy of yesterday's undemocratic technology decisions. However, a complementary, forward-looking effort has hardly begun to incorporate participation into the ongoing technology RD&D, investment, and policy decisions that will shape tomorrow.

In light of the broad reorganization of government technology policies and institutions underway in the aftermath of the Cold war's demise, coupled with the need to develop more democratically responsive institutions, there is an urgent need for comprehensive national debate and deliberation on science and technology policy. Appropriate initial venues could include televised hearings before key congressional committees or a televised national summit, followed by a pilot series of deliberative citizen tribunals (also summarized on television). A summit might be modeled on the informative economic summit that the Clinton transition team organized in December 1992, but with broader representation of alternative views.

8. Supportive Institutions

Previous steps propose a wide variety of new societal activities. Do local communities and governments have the competence and resources needed to do their part? A partial answer to this concern would be to redeploy existing, decentralized societal capabilities in a way that will enhance the efficacy of civil society—that is, of communities and secondary associations—rather than to fall back on the familiar polar alternatives of atomistic market processes versus expanded central government bureaucracy.[99]

Publicly financed or subsidized societal capabilities that can potentially be redeployed in support of technological democratization include universities, national laboratories, agricultural and manufacturing extension centers, industrial research consortia, and appropriate nonprofit organizations. These could be reorganized to constitute a national network of Community Research, Policy, and Assistance Centers. (Here the word "community" is intended to denote activities undertaken both for and collaboratively with communities—e.g., using participatory action research methods.)

A thriving model along such lines evolved in the Netherlands during the past 20 years. Dutch universities have established a network of public "science shops" to respond to the concerns of community groups, public-interest organizations, and trade unions about social and technological issues. Each shop's paid staff and student interns screen questions and refer challenging problems to university volunteers. During the shops' formative years, faculty members generally performed the research, but now graduate and undergraduate students do much of the work, under faculty supervision. Students who participate often receive university credit, and sometimes turn their investigations into graduate theses or adjust their career plans. Because students are doing research and writing papers, and faculty members are supervising and evaluating their work, both groups are doing what they would be doing as part of their regular workloads; thus the extra cost and time are minimal.

Incoming questions are generally accepted provided that the inquiring group lacks the resources to pay for research, is not commercially motivated, and would be able to use the research results as a basis for action. Some science shops also accept socially oriented inquiries from organizations—such as national environmental groups or local governments—that can contribute to the cost of undertaking research.

This system has helped workers evaluate the employment consequences of new production processes, social workers better understand the life circumstances of disaffected teenagers, and women assess the market potential for an independent women's radio station. In 1990, one of Amsterdam University's science shops branched out to undertake a study of air pollution on behalf of environmentalists in the severely polluted city of Dorog, Hungary. Sometimes a shop helps an inquiring organization conduct its own research.[100] Over time many of the science shops have specialized and now direct clients to the center best suited to addressing their concerns.

Today each of the Netherlands' 13 universities has between one and 10 science shops. Together the nation's 50 shops respond to several thousand inquiries per year. As a result of their work with science shops, some professors have conducted follow-up research projects, published scholarly articles in new areas, developed innovative research methods, forged new interdisciplinary collaborations, or modified the courses they teach. Thus the Dutch university system has to some degree been influenced to more directly serve society. Influenced by the Dutch model, other European nations, including Austria and Germany, now have their own science shops.[101]

The United States has no network comparable to the Dutch system. (In the late 1970s, the U.S. National Science Foundation began supporting similar activities through its Science for the Citizen Program. But the program was terminated abruptly early in the Reagan administration.[102]) Nevertheless, across America there are many individual examples of science shop–like activities that could function as initiating nuclei for developing a network. Besides those described in preceding steps, some others include:

- Chicago's nonprofit Center for Neighborhood Technology, which strives to devolve large technological systems and capital flows into systems that are more locally controlled and environmentally sound and that create productive jobs. Several years ago the center helped Chicago's metal-finishing industry devise a plan to detoxify and recycle its waste streams. This enabled a number of companies, each with strong employment records in low-income neighborhoods, to comply with environmental regulations and stay in business.[103]
- The University of Wisconsin's School for Workers, which is cur-

rently advising the national labor–management committee of the U.S. custom woodworking industry on approaches for including workers in designing and applying computer-aided production methods. The aim is to enhance productivity without compromising worker skills, wages, or safety.[104]
- Scientists at Brookhaven National Laboratory, who are involved in several community outreach efforts. One project is investigating methods for reclaiming Brooklyn's severely polluted Gowanus Canal so that the surrounding area can be returned to productive economic use. The techniques being developed could be of use in other difficult environmental cleanup efforts.[105]

Such examples suggest that organizing Community Research, Policy, and Assistance Centers (CoRPACs) can begin on an ad hoc local or statewide basis, pending federal support for establishing a nationwide network. Each center should have an advisory or governing board that includes empowered representatives of the communities and constituencies the center intends to serve.

At a later stage, further public accountability could be secured by channeling government funds through competitive grants.[106] As in the Netherlands, individual centers might specialize in serving particular functions or constituencies, while networking with other centers to share information, undertake collaborative projects, and direct client group inquiries to centers best suited to working with them. Among the many functions a national CoRPAC network might perform would be:

- Assisting local civic technological empowerment; defense conversion; and more democratic, self-reliant, and environmentally sustainable regional economic development.
- Facilitating worker, citizen, minority group, public-interest group, and local government participation in technology policymaking and in RD&D; and supporting constructive discussions and collaborations between the foregoing groups and the business community.[107]
- Contracting to produce participatory Social and Political Impact Statements, conducting supporting social trials, and assisting in integrative technology assessment.
- Establishing a national clearinghouse to help citizens and organizations locate pertinent information, government programs, and participatory opportunities, and to help prevent governments, businesses, and individual CoRPACs from duplicating one another's efforts (e.g., producing redundant social impact statements).
- Evaluating, comparing, and publicizing alternative policies, institutions, and experiences (domestic, foreign, and international) for advancing a democratic politics of technology.

- Helping primary and secondary school teachers learn the fundamentals of sociotechnological literacy and involve their classes in socially engaged projects.

In the case of universities, establishing CoRPACs would provide a healthy counterweight to professors' deepening ties to industry, which academe's fiscal duress and government policy both encourage. By consistently engaging community concerns, centers would help universities preserve their capacity for independent social criticism and educate students, via internships or role modeling and volunteer work, for responsible citizenship. (From a more self-interested perspective, universities would be apt to find that more directly serving communities is an excellent way to increase popular support for higher education.)

The end of the Cold war has thrown into doubt the future of the national weapons laboratories financed by the Departments of Energy and Defense, drawing attention also to the social functions and ad hoc evolution of the entire complex of national laboratories. (Including nonweapons labs, the United States has more than 700 national laboratories operating on a federal budget, in 1992, of $25 billion—which was one-third of the entire federal R&D budget.) A number of bills have been introduced in Congress that would shrink the national laboratory system, sell some of the labs off to the private sector, or refocus the weapons labs more toward commercial or other civilian missions.[108] Given their massive consumption of public funds, it is difficult to understand why national laboratories should not more directly and systematically serve community needs by participating in a CoRPAC network. Indeed, with time the CoRPAC system could evolve into the decentralized core of an alternative, strongly democratic, post–Cold war national laboratory system.

The federal government is currently subsidizing new industrial R&D consortia modeled after Sematech, the semiconductor industry's consortium based in Austin, Texas. But a coalition of local and national environmental, labor, and community organizations argues that Sematech itself has been insensitive to occupational and environmental hazards and has hired few employees from the poor and largely Latino community that surrounds its headquarters. Rebuffed in their attempts to influence consortium policies directly, the coalition helped persuade Congress to earmark $10 million of Sematech's 1993 budget for environmental, health, and safety R&D.[109] Subsidized heavily with public tax dollars, industrial research consortia, too, should become part of a national community-research network.

The Clinton administration is establishing a national network of manufacturing extension centers, intended to help small and medium-sized firms to adopt productivity-enhancing technologies and management practices. However, alternative methods of promoting productivity can either advance or hinder other social objectives, such as environmental sustainability or high-quality jobs. To help ensure desirable outcomes, manufacturing and agricultural extension services need strong worker, farmer, community, and public-interest group representation in their overall management and outreach programs. Integrated into a broader CoRPAC network, extension centers could, conversely, help nonprofit, community, and worker organizations adopt more effective management and coordination methods. Care is also needed to ensure that extension programs do not target individual firms at the expense of facilitating socially and economically fruitful relations among networks of firms, organizations, and communities.

The National Institute of Standards and Technology is creating an electronic communications link among federally financed manufacturing extension centers, and agricultural extension services are also becoming well linked electronically. These linkages could be integrated and enlarged to form an efficient communications and coordination backbone for the entire CoRPAC network. (If a strongly democratic variant of a National Information Infrastructure [NII] comes into being, that could much facilitate establishing and operating a CoRPAC network. On the other hand, if a CoRPAC network begins to emerge first, it would provide an obvious bottom-up venue for promoting a more strongly democratic NII.)

a. Paying for Democracy

Part of the cost of operating a CoRPAC network, paying citizens and experts for their public service, and other steps toward institutionalizing a democratic politics of technology could be recouped via a modest assessment against R&D or technological investment expenditures. A precedent exists in the budget of the federal Human Genome Project, of which 3 percent is designated for studies of the program's social implications. (R&D tax credits ought to help deter firms from concealing their R&D expenditures to avoid paying their share.) The net social cost should not, in any case, be exorbitant, insofar as many envisioned functions would be performed by reorganizing and redeploying existing institutions rather than by creating new ones. At the current level of U.S. federal and corporate R&D expenditure, a 1 percent levy would generate $1.5 billion annually— a respectable initial amount for launching a CoRPAC network. Potential sources for other needed funds are discussed in Steps 10 and 11.

b. The End of Innovation?

The concern sometimes arises that a social impact statement written 30 years ago about, say, computers would probably have failed to anticipate personal computers, word processing, and computer networks. Suppose it had highlighted only the dangers of centralized government surveillance and the destruction of jobs by automation? Would all further computer development have been banned? Fears of this sort are understandable. However, I believe that the kinds of institutions sketched in earlier steps ought to allay them.

Remember the uses suggested for impact statements. Political decisions regarding use of a new technology ought generally to be made locally, not nationally. Thus if some communities decided to guide, hinder, or even outlaw adoption of a particular technology, others would not. If the technology subsequently proved benign, the first communities could change their minds. A cost might be entailed, but it wouldn't mean the end of the world; and in the aggregate it should be more than offset by the social costs avoided through learning to adopt technologies more prudently.

Second, impact statements or political deliberations at earlier stages in a technology's development should generally enhance opportunities to create better designs—designs that will therefore be more favorably received. A representative of the Danish Council of Industry relates, for instance, that Danish corporations have benefitted from their nation's participatory approach to technology assessment because

> Danish product developers have worked in a more critical environment, thus being able to forecast some of the negative reactions and improve their products in the early phase.[110]

Finally, questions of the form "What if we had foolishly outlawed computers?"—or outlawed another contemporary technology—sometimes assume that we inhabit the best of all possible worlds and dare not tamper with the institutions that brought us this far. In reality, today's world is only one of an infinite number of others that we and previous generations might have fashioned instead. In creating this one, innumerable alternative technologies *were* overlooked, and at times actively suppressed.[111] Thus it is important to compare a realistically drawn democratic future against the real present, real past, and realistic projections of business-as-usual trends into the future—not against idealized caricatures.

A democratic politics of technology will certainly make mistakes. That is humanly inevitable. But it will also much enhance popular opportunities to engender a more socially responsive technological order while offering engineers and other professionals the satisfaction of knowing that they are helping to fulfill real social needs and aspirations.

SUPPORTING MACROCONDITIONS

The remaining several steps propose developing macrolevel institutions and social forces more consonant with democratic norms, thus—among other things—permitting fuller development of the preceding core activities. The creation of democratic macroconditions need not be done top-down; prior steps and bottom-up political struggles can contribute to evolving these macroconditions (recall Figure 12-1).

9. Corporations, Bureaucracy, and the State

Ronald Reagan rode to the presidency in 1980 promising to "get government off our backs." In practice, that slogan often amounted to a recipe for increasing the political–economic hegemony of business and the ultrarich. (According to the Congressional Budget Office, a startling 60 percent of all the growth in Americans' after-tax income from 1977 to 1989 went to the wealthiest 1 percent of families; meanwhile, real income among the bottom 40 percent of families declined.[112]) Might an alternative, strong democratic slogan be to "get government and the economy back into our hands"?

This implies striving for incentives or, eventually, mandating that corporations implement transitions to greater worker self-management, including worker participation in R&D and strategic planning. Legislation analogous to Scandinavian codetermination laws could require substantial worker, community, and public-interest group representation on corporate boards and on important board committees. Workers need empowerment particularly in decisions concerning their work tasks and work environment, while communities need influence on corporate decisions that will have significant, spillover local impact. Congress could enact a tax penalty against corporations that fail to democratize at a reasonable pace or that persist in producing democratically harmful technologies. New financial incentives might, if necessary, be made available to assist the startup of new businesses that are democratically favorable.[113]

Concerted efforts are also needed to democratize bureaucracy, both private and public. Potential strategies include legislative inducements to experiment with more socially permeable, decentralized, and democratically self-managed bureaucratic structures; devolving large organizations into loosely coupled assemblages of smaller ones; and increasing public responsiveness via citizen advisory boards and civil servant flextime and sabbaticals.[114] Moreover, to the extent that local communities become more democratically self-governed and more attentive to their effects on other polities—and that corporations are democratized—it should increasingly become feasible to prune back state and federal bureaucracies and to reorganize local-state-federal relations along more decentrally federated lines.

Historically, technologies of internal coordination and control have helped the United States become burdened with centralized, hierarchic organizational structures. Remedial programs of hardware and social science R&D in support of more democratic organizational forms are in order. For instance, personal computers and telecommunications harbor the potential for allowing more democratic, decentralized, and debureaucratized social coordination, but much remains to be learned about effective strategies for realizing that potential. As a recent business management study of information systems in 25 companies puts it:

> Information technology was supposed to stimulate information flow and eliminate hierarchy. It has had just the opposite effect.[115]

Regulations and effective enforcement mechanisms are also needed to ensure, for example, that solitary telecommuting from home is a job option, not an imposed requirement, to protect workers from computerized job pacing and surveillance, to protect consumers and citizens from invasive electronic surveillance of their shopping habits and political predilections, and to protect citizens generally from unwanted medical tests and genetic discrimination.

10. Military Technology, Democracy, and the Common Good

A thorough examination of the relationship between military enterprise, military technology, and democracy is beyond the scope of this work. Nevertheless, it cannot escape mention that war and military preparations have historically played a significant role in shaping and, at times, driving technological development.[116] The technological and social consequences are today felt throughout modern industry, in patterns of regional economic development, in the education of scientists and engineers, and in everyday consumer products and leisure activities (ranging from graphite tennis rackets to Rambo toys). Modern U.S. government institutions and policies for science and technology have been substantially a product of World War II's mobilization of science and of subsequent Cold war military preparedness.

A series of democratic complaints has been lodged against what President Eisenhower in 1960 called the "the military–industrial complex" and against the accompanying system of government that some label the "national security state"—a core domain of government relatively unchecked by democratic law and accountability. Political scientists warn that hair-trigger deployment of nuclear weapons subverts the constitutional balance of power between the executive branch and Congress. Other analysts argue that military R&D has contributed to forms of civilian industrial technology and organization that degrade work. Military secrecy in the corridors of government, within national weapons laboratories, and among defense contractors unquestionably poses a formidable barrier to democratic oversight, including to citizen participation in RD&D.[117] Recent press reports disclosing a hidden history of irresponsible military management of toxic wastes and harmful medical and radiological experiments performed on uninformed human subjects underscore dramatically the adverse repercussions of military secrecy. All of this sets aside the human and moral consequences of actually using weapons in war or exporting them for other nations' use.

The military also saps resources vital to technological democratization and to other urgent social needs. The government estimates that it will cost $300 billion over the next 30 years just to clean up hazardous waste at former nuclear weapons plants. For fiscal year 1995 the Clinton administration proposed spending more than 18 percent of the federal budget on defense—some $284 billion dollars. Of a total of $73 billion that the administration proposed to spend on R&D, 56 percent was allocated to military projects. That is down from the Reagan–Bush high point of 67 percent in 1988, but not yet even par with the 45–50 percent levels that prevailed during the 1960s and 1970s, even while the Vietnam War raged.[118]

Prospects for subjecting military expenditure and technology to critical democratic control have obviously improved with the end of the Cold war. But much remains to be done. Regional economic dependence on military expenditure has induced elected officials to support many weapons projects of dubious merit. For instance, senior arms control analysts insist that no conceivable enemy warrants the administration's proposal to deploy a new $25 billion antiballistic missile system—much less the more ambitious system favored by some congressional Republicans.[119] According to a 1994 *New York Times* editorial,

> The Pentagon has plenty of [weapons] programs to cancel, starting with Milstar, the communications satellites designed for fighting a nuclear war; the overpriced and underperforming C-17 transport plane; the multi-role fighter, and a new aircraft carrier. It can also cut back purchases of new aircraft like the F-22 stealth fighter.[120]

It would be far more wise and just to use such money to meet real social needs. The essential steps include, first, democratically defining realistic national security requirements; endeavoring via foreign policy to reduce the need to maintain armed forces still further; and striving to meet real security needs as effectively, democratically, and humanely as is possible. Indeed, insofar as strong democracy can be shown to contribute to reducing international insecurity and bellicosity, there would be instrumental security grounds to divert an increment of national security expenditure directly into advancing strong democracy both at home and abroad, including by democratizing technology.[121]

Second, military capabilities that are not needed should be converted to democratic civic or commercial purposes when that is feasible or else permitted to die with dignity and compassion. (It is not clear whether, in general, those defense contractors and national laboratories that have long histories of social and economic insularity and a preoccupation with weaponry can convert successfully.) The Clinton administration proposed spending $4.9 billion on defense conversion in fiscal year 1995. However, half of this amount was to be spent by the Pentagon rather than civilian agencies—mostly on dual-use technology development under the Technology Reinvestment Program (TRP).[122] ("Dual use" refers to technology developed for commercial application that can also be applied in advanced weaponry.) Insofar as dual-use funding is only made available for projects that have potential to serve military needs—and community stabilization and job creation and retention are not part of the program mandate—this is not really a defense conversion program at all.

Some members of Congress and their staffs worry that a dual-use strategy may actually permit the Pentagon to increase its penetration into the civilian economy at a time when just the opposite trend would seem to be called for.[123] Rather than taking it for granted that high levels of military R&D must be a perpetual given—a TRP premise—there should at least be a corresponding effort to negotiate stringent worldwide curbs or moratoria on weapons R&D (which might entail international exchanges of scientists and engineers to help verify that significant programs are not secretly underway and also cooperative international R&D on nonprovocative defense options).[124]

a. Jobs and Other Social Needs

Third, budgetary savings need to be invested toward meeting urgent societal needs (again, including technological democratization). High on the list of such needs are effective programs of support for local and regional conversion planning, job retraining, interim income support, relocation assistance, and stimulation or creation of high-quality jobs that satisfy further unmet social needs.[125] In the context of military downsizing, it is particularly important to ensure good jobs for groups traditionally subjected to adverse social discrimination, because one of the military's redeeming social accomplishments has been to break down unjust barriers to upward social mobility.

High-wage jobs are currently being squeezed and shed throughout the U.S. economy from many directions, including not only reduced military outlays, but also Wall Street's faddish infatuation with corporate downsizing, more efficient management strategies, new labor-displacing technologies (which current federal technology programs may accelerate), and export of jobs abroad (which is facilitated by new international telecommunications networks). On the other hand, there is no lack of areas of great social need able, in principle, to provide meaningful work: ecological restoration and stewardship, social work, education, preventive medicine, sustainable agriculture, energy conservation, designing and building democratic infrastructure, improving and expanding convivial public spaces, and so on.[126] The key is to seek ways of redeploying socially unproductive military resources, as well as idle or underemployed human creativity, so that they can contribute to satisfying these or other democratically established needs.

Resources invested this way could still be used to maintain and expand regional economies and jobs, but the jobs would be serving communities and the common good rather than marking time or, worse, contributing to international arms sales, arms races, and the miseries of war and oppression. Moreover, insofar as military production tends to be unusually capital-intensive, conversion to socially useful production could be a vastly more efficient job generator.

Genuine defense conversion and technological democratization would interact synergistically to their mutual benefit. For instance, access to Community Research, Policy, and Assistance Centers would facilitate all aspects of conversion planning and implementation, and civic technological empowerment involves many steps vital to community stabilization and job creation. Conversely, conversion planning can be an excellent vehicle for starting to plan and implement all aspects of civic technological empowerment.

11. Transnational Corporations, Foreign Policy, and the World Economy

Are certain forms of corporate enterprise (notably, large transnationals, international banks, and giant investment firms) structurally resistant to democratization—that is, are they too large to democratize internally and too powerful, individually and in combination, to subordinate to external democratic guidance? How can one identify such corporate forms and seek their devolution toward democratically manageable alternatives? A first step might be to establish a popularly based international study commission, perhaps organized as a satellite-televised citizen tribunal. At the national level, one instrument would be to enlarge the justification for government antitrust activity to encompass the concern that large corporations acquire unjustified political and economic power. It is also crucial to establish democratic regulation of international investment and capital mobility.[127]

All aspects of U.S. foreign policy and involvement in international institutions ought to be reviewed with an eye toward their domestic and foreign structural ramifications for democracy.[128] How do United States–based multinational corporations, U.S. agencies such as the Export–Import Bank, and multilateral lending institutions affect developing nations' capacities to pursue more self-reliant and participatory development paths?[129] For instance, the U.S. Agency for International Development and analogous agencies in other nations may impair developing nations' self-reliance by providing assistance contingent on recipients' agreement to import goods, services, or technology from corporations based in the donor nation. How ought international trade accords—especially the General Agreement on Tariffs and Trade (GATT)—to be amended or supplemented to allow fully effective policies for local self-reliance, ecological sustainability, and other democratic objectives?[130] How can communities and everyday citizens from all over the world begin to achieve democratic representation in the negotiations and institutions that regulate world trade, the transnational division and organization of labor, the evolution of global communication and transportation systems, the control of indigenous knowledge and of biological and ecological resources, and in other dimensions of international and transnational relations, including R&D and investment?[131] There are recent signs of hope in the unprecedented levels of involvement of grassroots-based organizations and social movements in global environmental conferences and in political disputes over the GATT and the North American Free Trade Agreement (NAFTA).

Are new international or transnational institutions needed to support a more democratic world technological order? One option would be a United Nations–based Peace Corps or other vehicles for internationalizing citizen sabbaticals; this seems particularly sensible given some history of bilateral aid advancing donor interests at the expense of recipients. Other possibilities include a U.N. Office of Technology Assessment to facilitate production of transnational social and environmental impact statements and technology policy studies; integrating the proposed CoRPAC network (Step 8) into an analogous transnational network; and regional participatory RD&D centers oriented toward a more locally self-reliant, environmentally sustainable, and democratic world order.

There is probably need, too, for one or more transnational organizations to nurture communications systems that would reflect more pluralism and minority coverage in international news coverage and support transnational democratic dialogue among communities and social movements. (While there is some hope for emerging international information systems to facilitate transnational grassroots political deliberation and coordination,[132] it seems obvious that for the time being the world's affluent professionals will be disproportionately active in global electronic communications networks.)

Chapter 7 described numerous structural democratic benefits associated with more locally self-reliant economies. Transnational corporate activity and world trade, on the other hand, have tended to subvert local self-reliance and democratic self-determination. Thus, from a societal point of view, some of the high productivity and scale economies conventionally attributed to transnational production amount, in terms of neglected first-order structural harms (or benefits foreclosed), to extensive false economies. Given that the international flow of money in banking, investment, and currency speculation is now at least several hundred times greater than the value of traded goods and services,[133] a partial remedy would be to seek a modest global tax—perhaps just a few tenths of a percent—on global electronic financial and investment flows. This money could be rebated to subnational governments, nonprofit organizations, community-oriented businesses, and so forth to provide seed capital for the transition to more sustainable and self-reliant local economies.

More diversified local production for local consumption has the potential to provide high spillover social benefits and to create large numbers of meaningful new jobs, but not necessarily at wages equal to those currently provided in transnational corporate employment. Some of the reduced wages may be offset naturally in the form of direct psychological satisfactions as well as sundry social amenities and goods provided through life in a more democratic local community.[134] To the extent, however, that transnational corporate and financial activity is now associated with spawning unemployment, underemployment, and other significant social and democratic structural costs—and is often subsidized via extensive public investment in supporting infrastructure—an additional global tax for use in subsidizing incomes in the self-reliant sector would, in principle, be warranted.

As one example of the aforementioned, democratic structural costs: During the first two years of the Clinton administration, the only issue on which the president went to the mat and absolutely would not back down had nothing to do with the bold domestic agenda on which he had campaigned for office. Rather, it was to secure congressional approval of the NAFTA and GATT trade-liberalization accords, both of which were vigorously opposed by key elements of his party's core constituency—organized labor and environmentalists. What can account for the president's surprising behavior? It is tempting to speculate that with global information technology now making possible massive and instantaneous nationwide disinvestment—witness the transnational capital flight and precipitious currency devaluation that Mexico suffered at the end of 1994—a president simply cannot afford to disregard the interests of the world financial community. One needn't postulate any conspiracy at work here. The outcome can be explained, instead, by the simple combination of technological capability, market pressures experienced by financiers, and political leaders' basic survival instincts. In short, the electronically enabled global financial system appears to be shaping central elements of the United States' political agenda—including what elected leaders must (and must not) say and do—severely compromising the democratic process.

Major political efforts are also required to guide the world economy toward more peaceful and equitable pursuits. Today something on the order of one-quarter of world R&D is directed toward the military, while all of the world's developing nations taken together—which include humanity's largest and neediest populations—perform perhaps 5 percent of total world R&D.[135] (Because the United States is responsible for about 40 percent of total world R&D expenditure, democratic reform of U.S. R&D policy would of itself have global implications.)

A more democratic foreign policy need not depend entirely on federal involvement. Citizens and communities have already taken initiatives in establishing democratic transnational alliances via nongovernmental environmental, peace, labor, human rights, and appropriate technology organizations and via municipal foreign policies.[136]

CONCLUSION

This chapter offers preliminary suggestions for debate—certainly enough to keep a healthy conversation going. The vision is ambitious but hardly overwhelming or unrealistic. As the chapter's examples attest, many pertinent subsidiary policies and institutions already exist and thus only require expansion, replication, or some redirection or modification, not invention from scratch. Countless individuals and organizations are already hard at work. An ancient Chinese proverb reminds us that the journey of 1,000 miles begins with the first step, and we are already much further along than that.

Chapter 13

"A NEW AND BETTER VISION"

Although my argument is contestable, several of its underlying insights seem relatively robust. It is fruitful to interpret technologies as social structures. Because they cannot comprehend technology's public, structural role, both conventional economic analysis and unregulated markets are seriously inadequate for guiding technological development and policy. Of course, the contemporary politics of technology also slights technology's structural role; technology is not the only factor impairing democracy, but it is as important as any other and is the least understood. However, one can anticipate constructive lay involvement partly because esoteric technical knowledge is generally not central to grasping technologies' nonfocal, structural aspects. Are these insights, taken together, not sufficient to warrant seeking some sort of democratic politics for engendering democratic technologies?[1]

From the standpoint of an envisioned strongly democratic politics of technology, one can interpret contemporary technological politics in terms of several components. Many of the most important technology decisions are made today via a covert politics that occurs within corporate headquarters and government bureaucracies or via the tacit politics of the economic marketplace.[2] Meanwhile, one might characterize the overt, publicly enacted politics of technology as roughly comprising the wrong people posing the wrong questions about the wrong technologies at the wrong time.

In recent decades public controversies have engulfed many technologies (e.g., the fluoridation of public water supplies, electric power plants, advanced weapons systems, hazardous waste disposal methods, telecommunications security systems, etc.). Time and again powerful political actors (e.g., professional politicians, government administrators, corporate leaders, and representatives of major organized interest groups), complemented by competing teams of technical experts, have ritualistically debated the economic costs and consequences, military security implications, health and safety risks, or environmental impacts of the current crop of topical technological developments. The addressed questions are not inconsequential, but this system of public discourse and decision making is nevertheless inadequate because it:

1. Excludes lay citizens from anything but a trivial role;
2. Normally raises questions only long after many of the most important decisions (e.g., over the available range of technological alternatives) have already been made elsewhere;
3. Evaluates technologies almost exclusively on a case-by-case basis, discouraging action on, or even identification of, socially significant complementarities among focally unrelated technological developments;
4. Focuses on just a few "cutting-edge" technologies to the virtual exclusion of the vast majority of emerging and, even more so, existing technologies;
5. Directs attention primarily toward these publicized technologies' promised material, focal purposes at the expense of often more important cultural and other nonfocal social consequences; and
6. Leaves the question of the debated technologies' structural bearing on democracy as only the most important of many questions that are never asked.

That encapsulates the procedural side of contemporary technological politics. What of the substantive nature of our technological order? Have undemocratic decision-making processes at least produced creditable results? A century and a half ago Alexis de Tocqueville described a politically exuberant United States in which steaming locomotives could not restrain citizens' ceaseless involvement in politics and community life:

> In some countries the inhabitants seem unwilling to avail themselves of the political privileges which the law gives them; it would seem that they set too high a value upon their time to spend it on the interests of the community; and they shut themselves up in a narrow selfishness. . . . But if an American were condemned to confine his activity to his own affairs, he would be robbed of one half of his existence; he would feel an immense void in the life which he is accustomed to lead, and his wretchedness would be unbearable.[3]

That is not today's United States, in which a bare majority of eligible voters participate in presidential elections while usually even fewer engage in local politics or in nonelectoral issue-oriented national politics. A study issued in 1991 by the Kettering Foundation reports that Americans yearn to be more involved in public affairs but feel locked out of consequential decision-making arenas.[4] The causes of citizens' disaffection and political disengagement are complex, but Part II suggests that technology has played a much more significant and intricate role than is commonly believed.

Current technological orders are generally short on communitarian or cooperative activities but long on isolation and authoritarianism (violating Figure 5–1's Criterion A). Work is frequently stultifying or harshly stressful, tending to impair moral growth, political efficacy, and self-esteem (violating Criterion B). Illegitimate asymmetries in social power are reproduced through clandestine technological means, including ideological means (violating Criteria C and D).

The opportunity to engage in a vibrant civic life and in its accompanying informal politics is often preempted by such modern technological complexes as shopping malls, suburban subdivisions, unconstrained automobilization, and an explosive proliferation in home entertainment devices. Thus people have diminished access to local mediating institutions or to public spaces that could support democratic empowerment within the broader society (violating Criteria A and G). The need to manage translocal harms, coupled with widespread dependence on large, centrally managed technological systems and growing local integration into global markets, has helped render local governments relatively powerless, thereby reducing everyone's incentive to participate at the local level (violating Criteria E, F, and G). Meanwhile, there is little compensating incentive to engage in national politics, which network television reduces to a passive spectator sport, where hyperempowered corporations exert disproportionate influence, where deep questions of social structure are slighted, and where the average citizen has negligible effect.

While it is not always easy to establish causal connections running from structural deficiencies to other social ills, it seems inconceivable that weak community ties, atrophied local political capabilities, and authoritarian and degraded work processes have had no influence on growing rates of illiteracy, stress, illness, divorce, teen pregnancy, crime, drug abuse, psychological disorders, and so on.[5] Perhaps, as de Tocqueville foresaw, many people do sometimes feel shut up in a narrow selfishness, robbed of one half of their existence, and left with an immense void in their lives.

Left to its own devices, the world is not going to get better. As in Ibieca, the nonfocal nature of technologically embedded structural change allows degeneration to advance insidiously. Standards of democracy, personal freedom, and well-being can erode in small steps and degrees, not with a bang but a whimper. Thus people do not have the luxury of adapting to the status quo, because the status quo continues to deteriorate.

However, there are notable exceptions to these dismal generalizations. For instance, the Dutch science shops and Denmark's consensus conferences are two among many indications that Western Europe is already outcompeting the United States in supporting the involvement of laypeople in technology decision making. Likewise, some of the best examples of participatory research are being implemented today in the developing world.[6]

But the end of the Cold war has elicited some promising indications of change in the United States too. At the national level, a number of government agencies have begun to explore some of the components of a more democratic politics of technology. For instance, in 1994 the National Institute of Standards and Technology sponsored a national conference and regional meetings on worker participation in technology development.

Congressman George E. Brown, Jr. is widely acknowledged as one of the most thoughtful and influential legislators on U.S. science and technology policy; he is also a strong proponent of a more democratic and socially responsive technological order. In April 1993 Brown addressed the annual colloquium on science and technology policy organized by the American Association for the Advancement of Science. His speech distinguished today's "market-driven" technological change from a "socially oriented technology policy" in which markets and technological development would be subservient to fundamental human goals as expressed "through a conscious process of democratic action":

> A social–technology policy might actually proscribe some types of technological solutions. . . . A social–technology policy might encourage the development of technologies that decentralize political power and economic resources, to minimize the control that large institutions have over the lives of individuals.
>
> For the past 50 years this nation has focused its resources on building weapons of inconceivable destructive power, and we have viewed the rest of the world as a chess board. . . . We provided high-technology weaponry to dictatorships. . . . Our vision during the Cold war was cynical in the extreme.
>
> Now the Cold war is over, and our excuse for this behavior is gone. We need a new and better vision. . . . Is our path into the future

to be defined by the literally mindless process of technological evolution and economic expansion, or by a conscious adoption of guiding moral precepts?[7]

When Brown concluded, the august assemblage of research leaders and administrators that had gathered in the nation's capitol rose to give him a standing ovation.

Of course, programs, people, and passion can come and go at the national level. For instance, when Brown made the preceding remarks he was the chairman of the House Science Committee; but with the Republican party's November 1994 congressional electoral victories, he became that committee's ranking minority member. In contrast, ferment throughout lower layers of the social fabric may indicate more pervasive change or opportunity. There one can find a growing number of democratic workplaces, cohousing projects, technologically disadvantaged groups mobilized for redress, associations of technical experts collaborating with grassroots groups, nonpaternalistic and holistic approaches to health care, ecologically sensitive businesses and farms, nonprofit organizations that promote community-based telecommunications systems, communities that are crafting vibrant public centers, local governments that seek to enhance economic self-reliance, transnational sister-cities projects, and broad social coalitions supporting a more sustainable, just, and demilitarized economy. This book has striven to supply a conceptual framework binding such examples together, demonstrating that many rudiments necessary for a more democratic politics of technology already exist. (Moreover, if community solidarity has weakened, individuals have also escaped some previously oppressive variants of community integration. That at least affords some opportunity to try to rebuild community along democratically more satisfactory lines.)

Is a democratic politics of technology feasible? There is already a highly evolved version operating at small scale among the Amish. There are also less evolved but larger scale examples. In various nations, especially in Scandinavia, trade unions have begun securing the right to participate in developing more democratic workplace technologies. In the United States, there are at least two major examples in which large numbers of laypeople are already involved in evaluating entire technological orders' nonfocal effects, using nonmarket democratic design criteria, and sometimes even participating directly in technological research or design. One is the movement among people with physical disabilities for barrier-free design; the other is the environmental movement. Once the number of democratic criteria broadens and the range of participants expands, we will be well on our way toward a healthy capacity to generate a more democratic technological order.

This book has a simple message: it is possible to evolve societies in which people live in greater freedom, exert greater influence on their circumstances, and experience greater dignity, self-esteem, purpose, and well-being. The route to such a society must include struggles toward democratic institutions for evolving a more democratic technological order. Is it realistic to envision a democratic politics of technology? Isn't it unrealistic not to?

NOTES

Notes to Chapter 1

1. Mumford (1964, p. 7).
2. Harding (1984).
3. See, for example, CISRD (1992) and Branscomb (1993).
4. I use the word "technology" broadly to encompass material artifacts (including buildings) and the social processes and knowledge accompanying their development and use. I use "technologies" (plural) to connote the disparate social effects associated with different artifacts and practices.
5. The technological dimensions of some of these transformations are discussed in Colton and Bruchey (1987).
6. See also Winner (1986, p. 10).
7. Olshan (1981).
8. See Olshan (1981) and Kraybill (1989). Eric Brende's generous readiness to share his essays-in-progress and his personal experience living with an Amish family also greatly helped my understanding of Amish technological decision making.
9. Amish community decisions are made collectively, normally following the weekly religious service. Religious–political leaders are nominated by adult community members and chosen by lot. Amish society is not fully egalitarian sexually; women participate politically but are ineligible to hold formal leadership positions. Adult membership in the community is voluntary, in the sense that one becomes an adult member upon choosing, during late adolescence, to undergo baptism into the Amish church. Earlier adolescence typically includes an informal period of communally tolerated "acting out," in which Amish teenagers may adopt non-Amish clothing and hairstyles, listen to popular music, drink alcohol, or purchase automobiles. Thus, the choice to become an adult member is informed by some experiential knowledge of alternative lifestyles in the non-Amish world. Today, upward of 80 percent of all Old Order Amish children choose to remain Amish as adults (Kraybill 1989).
10. Olshan (1981, p. 303).

Notes to Chapter 2

1. Winner (1986, p. 29).
2. The story of Ibieca is not unique. I will refer to it repeatedly because it vividly illustrates some general lessons concerning technology's social dimensions.
3. Jackson (1985).
4. For illustrative examples, see Part II, especially Chapters 4 and 5.
5. Carroll (1977, pp. 338–339).
6. Sommer (1969).
7. Giddens (1979, pp. 69–70).
8. Zuboff (1988).
9. In economic theory these are impacts imposed on someone without the mediation of a perfect competitive market. See Chapters 7 and 10.
10. See Borgmann (1984).
11. Lee (1959, p. 31).
12. Turkle (1984).
13. Luria (1976) and Elias (1978).
14. White (1962).
15. Chandler (1977), Kasson (1977), and Stephens (1989).
16. Winner (1977; 1986, chap. 2).
17. See also Balabanian (1980).
18. On structures shaping, but not determining, social reality, see, for example, Unger (1987).
19. Schrank (1978, pp. 221–222).
20. Sclove (1982).
21. Bijker et al. (1987, pp. 42–44).
22. E.g., Goldsmith et al. (1976).
23. E.g., Lovins (1977).
24. See Callon (1980, 1987) and compare Pfaffenberger (1992).
25. See, in addition to the examples cited in the succeeding text, Piore and Sabel (1984), Bijker et al. (1987), and Feenberg (1991).
26. Cowan (1983, chap. 5).
27. Hayden (1984, chap. 4).
28. Smith (1970) and Pacey (1976).
29. Collingridge (1980), Hughes (1989), and Callon (1994, pp. 407–408). For a case where late-stage design change did take place, see the Dutch dam example in Chapter 8.
30. E.g., OTA (1987).
31. See, for example, Zimbalist (1979).
32. The example is from Professor Merrit Roe Smith's fall 1979 M.I.T. lectures on "Technology in America."
33. E.g., Falk (1984).
34. McLuhan (1964, chap. 1).
35. On the contingent historical development of privacy and individualism,

and its relation to material artifacts, see, for example, White (1974) and Elias (1978).

36. Winner (1977). One curious fact about these two prevalent myths is that not only is each false, but they are also mutually exclusive. That is, if technologies were indeed neutral, they could not also be an autonomous force that determines the course of history. While false, these myths are nevertheless socially consequential. Each helps deflect the critical political scrutiny that technologies warrant. The myth of neutrality does so by saying that artifacts themselves are irrelevant; one need only be concerned about the people who use them. The myth of autonomy does so by saying, in contrast, that technologies matter very much, but that people can have no choice in the matter; technologies themselves are running the show. By competing with one another, these myths comprise a unified ideological system that deflects attention from the relatively simple insight that technologies are contingent social structures.

37. Winner (1986).

Notes to Chapter 3

1. Tripp (1980, pp. 19, 33).
2. Barber (1984). I find the term "strong democracy" congenial and I agree with many of Barber's ideas. However, I also draw on other theoretical treatments that complement or qualify his model, such as Rawls (1971), Mansbridge (1980), Bowles and Gintis (1986), Gould (1988), Young (1989), and Cohen et al. (1992). These works also include extensive criticism of thin democracy or its underlying, classical liberal presuppositions; see also Chapter 10 (especially the concluding section).
3. See, for example, Barber (1974) and Evans and Boyte (1986).
4. Barber (1984). Readers favoring a contemporary sound are invited to substitute the label "Democracy Lite."
5. The contrast between the strong and thin democratic traditions is real, but it can also be overstated. In practice, democratic societies tend to exhibit a shifting mixture of strong versus thin practices.
6. Social struggles, the influence of business interests and social elites, and city planning have all, of course, left their mark on contemporary cities. But often planning has done little but help establish the infrastructure under which anarchic growth could proceed. On the technologically influenced evolution of American urban and suburban areas, see, for example, Mumford (1961), Cowan (1983, esp. chap. 4), Hayden (1984), Jackson (1985), and Rose and Tarr (1987).
7. Compare Barber (1984, esp. pp. xiv, 151).
8. Berger (1977, vol. I, p. 22).
9. Sources on the Berger Inquiry include Berger (1977), Gamble (1978), and OECD (1979, pp. 61–77).
10. Tripp (1980, p. 19).

11. E.g., Barber (1984, chap. 10) and Goldhaber (1986, part II).
12. E.g., Mander (1978).
13. See Part II.
14. On the art and social process of design, see Pye (1982), O'Cathain (1984), Ferguson (1992), and Thomas (1994, chap. 7). On democratic judgment generally, see Beiner (1983) and Barber (1985).
15. Sclove (1983, pp. 44–45), Schwartz (1984), and Barber (1985).
16. On the general concept of rationally contestable norms, see, for example, Bernstein (1983). I use the term "rational" broadly to connote the noncapriciousness that is realized via uncoerced deliberation and embodied afterward in substantively moral results—that is, in judgments or actions consistent with perpetuating freedom and democracy. Hence the phrase "democratically contestable" is a synonym.
17. Examples of conceptions of technology and democracy that are either purely procedural or purely substantive include, respectively, OECD (1979) and Mumford (1964). Notable prior steps toward synthesizing these two components include Illich (1973), Alexander et al. (1977), Bookchin (1982), Goldhaber (1986), and Feenberg (1991).
18. E.g., Ferguson (1979, pp. 14–18).
19. E.g., Gould (1988, chaps. 1 and 3).
20. Kant (1959, esp. pp. 46–54). The discussion that follows draws on selected Kantian and Rousseauian insights, but also differs in significant respects. For instance, I would strongly qualify Kant's designation of cognitive reason as the sovereign faculty of moral judgment and also the formulation of the categorical imperative that identifies morality with compliance with formal rules.
21. Ibid., p. 47. For evidence that the categorical imperative is not simply a particularistic expression of Western values, see, for example, Erikson (1964). Even modern critics of Kant often wind up endorsing ethical views not obviously far removed from the categorical imperative. For example, Rorty (1989) espouses a public ethic of minimizing cruelty, which—while weaker—is certainly consistent with the categorical imperative.
22. In its essential logic, this solution is Rousseau's in *The Social Contract* (1968). See also Cassirer (1963).
23. Barber (1984, pp. xiv, 151).
24. Young (1989).
25. E.g., Rawls (1971, pp. 73–89, 274–284, 298–303).
26. E.g., Barber (1984, pp. 290–293). On Athenian democracy, see Finley (1983, pp. 70–84).
27. See, for example, Mill (1972, esp. pp. 346–359), Dewey (1954, pp. 151–155, 211–219), Unger (1975, pp. 173–174, 261), and Mansbridge (1980, chap. 20).
28. See, for example, Kant (1970, pp. 90, 102–105), Alperovitz (1979), Horvat (1982), and—although some of her arguments for centralization are overstated—Rose-Ackerman (1992, chap. 11).
29. E.g., Frug (1980), Rae et al. (1981), Thompson (1983), Gould (1988), and Shuman and Harvey (1993, pp. 88–89). On transnational or trans-societal

moral obligations and political responsibilities, see, for example, Miller et al. (1988).

30. Compare Sirianni (1984) on the importance of institutional pluralism and also Bowles and Gintis's (1986) treatment of heterogeneous power. On associations and movements, see Cohen et al. (1992).

31. E.g., Cobb and Elder (1972).

32. E.g., Rawls (1971, p. 61).

33. Small communities can sometimes be parochial or oppressive, although—by definition—not if they are strongly democratic. For evidence and argument contrary to the view that local communities are invariably parochial or oppressive, see Mansbridge (1980), Taylor (1982), and Morgan (1984). Various social observers (e.g., Lee 1959, 1976) argue that the dangers of enforced social conformity are greater in a mass society than in one composed of smaller, differentiated, democratic social units. On communities as mediating structures that empower individuals, see Frug (1980), Bowles and Gintis (1986, chaps. 5 and 7), and Berry et al. (1993). The potential power inherent in local democratic community is made manifest, for example, in Old Order Amish people's success in resisting the joint coercive force toward conformity resulting from external market relations, persistent social ridicule, and one-time government opposition to various Amish practices (such as refusal to participate in Social Security, military service, and the public school system).

34. See Rawls (1971, sect. 35) and Gutmann (1987, pp. 41–47).

35. E.g., Turnbull (1972) and Erikson (1976).

36. E.g., Schwartz (1984) and Young (1989, pp. 262–263).

37. See, for example, Dewey (1954, pp. 211–219), Mansbridge (1980, pp. 301–302), Wolfe (1989, pp. 133, 233, 257–258), Barber (1992, esp. pp. 64–65), and the debate in Nussbaum et al. (1994).

38. See, for example, Calhoun (1991).

39. Brugmann (1989).

40. On the logic and psychology of generalizing from particulars, see Polanyi (1969, part III) and Cicourel (1981).

41. For one reasonable attempt at synthesizing a number of contrasting modern views on individual moral development, see Kegan (1982). Empirical research has not decisively confirmed or refuted the thesis that self-governed workplaces encourage political participation beyond the workplace (e.g., compare Elden 1981 with Greenberg 1986).

42. E.g., Schwartz (1982), Dahl (1985), and Gould (1988).

43. Schor (1992). For examples of contemporary workplaces in which menial tasks are shared, see Rothschild and Whitt (1986, pp. 108–111) and Holusha (1994). Feminist writings on the importance of sharing childcare responsibilities are also salient here (e.g., Hochschild 1989).

44. E.g., Sirianni (1988).

45. Lovell and Stiehm (1989, pp. 187–188).

46. For perspectives on democratic knowledge production and civic education, see, for example, Schutz (1946), Adams and Horton (1975), Feyerabend (1978, part II), Goodwyn (1978), Foucault (1980, chaps. 5–6), Freire (1980), Geuss (1981), Barber (1984, chaps. 8–10), Belenky

et al. (1986), Gutmann (1987), Marglin and Marglin (1990), and Harding (1993).

47. There are two, mutually contradictory objections not addressed here. Communitarians (including some Marxists, anarchists, and poststructuralists) may leap to the conclusion that strong democracy must be radically individualist if it has Kantian roots (e.g., Sandel 1982). Liberal individualists, certain feminists, and others may leap to the contrary conclusion that strong democracy must court totalitarianism if it has Rousseauian roots (e.g., Popper 1966; Young 1989). However, the text implicitly departs from Kant and Rousseau in a number of respects in order to accommodate reasonable criticisms along the preceding lines.

Some postmodernist thinkers may oppose strong democracy as a facet of their principled opposition to all prescriptive social theory or foundationalism. The text may partly allay such concerns by acknowledging its own contestability and favoring relatively malleable, egalitarianly self-governed social structures (see Part II). Moreover, it seems reasonable to ask those who simply deconstruct others' visions to answer the counter-criticism that (1) people do not have the option of living in a structureless world, while (2) worlds that are perpetually self-deconstructing may harbor deep problems of their own (see Chapter 8). Finally, those postmodernists who do muster the courage to prescribe seem to have difficulty not tacitly reinventing modern positions such as conventional political pluralism, liberalism, ideal speech situations, or even egalitarian ethics not far removed from those espoused here (e.g., Rorty 1989; Alcoff 1989, esp. pp. 323–324).

48. E.g., Lindblom (1977) and Bowles and Gintis (1986).

49. E.g., Schrank (1978) and Kohn et al. (1990).

50. E.g., Bowles and Gintis (1986) and Evans and Boyte (1986).

51. Kemeny (1980, pp. 73–75).

52. Young (1990, pp. 233–234); see also Dahl and Tufte (1973, p. 6). Young's characterization of strong democracy seems, incidentally, something of a caricature. Strong democracy, as envisioned here, involves greater local self-reliance but not total self-sufficiency (see Chapter 7), face-to-face communities but also nonterritorial groups and communities (see Chapters 4 and 6), and quasiautonomous polities within a broader federated system.

53. See also Chapter 10; and compare Rawls's (1971) concept of lexically ordered moral principles, with priority granted to equal liberty.

54. E.g., Prewitt (1983, pp. 51–52, 56).

55. Booth (1988).

56. Gamble (1978, pp. 950–951). For other examples of laypeople contributing to technical understanding, see, for example, Krimsky (1984) and Brown and Mikkelsen (1990).

57. E.g., Kemeny (1980) and Mazur (1986, p. 249).

58. E.g., OECD (1979), Petersen (1984), and *Technology and Democracy* (n.d.).

59. Prewitt (1983, p. 51).
60. See also Pacey (1983, pp. 155–159).
61. See also Brown and Mikkelsen (1990, esp. p. 146). For evidence that technical experts do not invariably oppose lay participation in technical decisions, see Lakshmanan (1990).
62. Case studies reviewed in Chapter 11 lend credence to this suggestion; so does Hays's (1969) history of the Progressive conservation movement.
63. Dutton et al. (1988, pp. 326–327) and Dickson (1988, pp. 232, 257–258).
64. Sclove (1983, p. 45). On the community of experts as self-interested political actors, see also *Health of Research* (1992, pp. 7–8).
65. Gamble (1978, pp. 950–951).
66. See Chapter 2.
67. Gray (1988).
68. Clinton and Gore (1994, p. 28).
69. Defenders of elite-dominated decision making sometimes erect the straw-man argument that lay participation can only mean deciding everything by public referendum, preceded by a manipulative media blitz (e.g., Mazur 1986, p. 249). On the other hand, Harvard professor Harvey Brooks (1984, p. 48) concedes "that when a representative 'jury' is deeply exposed to the issues . . . , they will usually make sensible decisions." But because U.S. referenda have often not incorporated the requisite educational opportunities, Brooks throws his weight instead toward expert decision making without considering obvious alternatives such as expanding the opportunities for the routine establishment of representative panels ("juries") of lay citizens (see Chapter 12).
70. See, for example, White (1974) and Corn (1986).
71. De Sola Pool (1983), Lowrance (1986, pp. 133–134), and Sclove (1989).
72. See Chapter 12, n83.
73. E.g., Nelkin (1992).
74. Some past errors in prediction might, for example, be lessened in the future by the following:
    1. learning to discount mere technical possibilities in favor of the consequences that are most probable in a specific social and technological context (White 1974; Sclove 1989);
    2. looking out for the emergence of a new technology's supporting background conditions; for the unintended coercive dynamics that can accompany technological innovation; or for the tendency of a new or maturing technology to elicit complementary technologies, while latently foreclosing alternative sociotechnological possibilities (Hughes 1989; Kraybill 1989, chap. 11; Sclove and Scheuer 1994; Rosenberg 1994, pp. 1–23, 205–223; and Chapter 10);
    3. scanning for stylistic complementarities among aggregates of focally unrelated technologies or for structural constants lurking amid a distracting sea of focal technological change (as in Ihde 1979; Cowan 1983; Borgmann 1984);

4. continuing to study past predictions to seek generalizations concerning promising approaches (Corn 1986; de Sola Pool 1983;
Sclove 1989); and

5. ensuring that all potentially affected groups are fully represented
in the prediction process, so that key questions, concerns, and
contextual knowledge are less likely to be overlooked.

75. E.g., Hughes (1983) and Bernard and Pelto (1987).

76. Perrin (1980); see also Noble (1979) and Zuboff (1988, chap. 10).

77. Olshan (1981) and Logsdon (1986).

78. Kraybill (1989, chaps. 7 and 11) and Stoltzfus (1973, p. 201).

Notes to Chapter 4

1. I use the compound "communitarian/cooperative" to connote two traditions: the communitarian tradition (which emphasizes community or
home life) and the cooperative tradition (which emphasizes work life).
The compound is intended to suggest that the same democratic considerations apply in all social domains (see Chapter 3), and also that there
may sometimes be democratic advantages in blurring the distinction
between various social domains.

2. Compare Rawls (1971, esp. pp. 107, 467) and Gould (1980, pp. 83–89,
116–178).

3. On the contingent social origin of hierarchy and its embodiment in
technological artifice and practice, see, for example, Mumford (1967), Du
Boff and Herman (1980), and Bookchin (1982).

4. De Kadt (1979, p. 244).

5. Boehmer and Palmer (1994).

6. Quoted in Zuboff (1988, p. 337).

7. Zuboff (1988).

8. See also Gintis (1972, esp. pp. 584–598). Workplace analysts (e.g., Sabel
1982) often overlook the role of nonworkplace technology in influencing
culture and behavior at work.

9. Cowan (1983).

10. Kramarae (1988).

11. Hochschild (1989).

12. E.g., Sorkin (1992). On air conditioning, see Jackson (1985, pp. 280–281)
and Hinton (1988). On central heating, see Borgmann (1984, chap. 9)
and White (1974, p. 364).

13. This criticism is less applicable to that facet of the appropriate technology movement oriented toward the developing world. On appropriate
technology generally, see, for example, Darrow and Saxenian (1986).

14. Winner (1986, chap. 4); for an important exception, see Dickson
(1974).

15. Best (1990), Hirst and Zeitlin (1991), Kanter (1991), and Holusha (1994).

16. Hayden (1984, pp. 33–34, 46, 146), McCamant and Durrett (1988, pp.

10, 25), Oldenburg (1989, chap. 11), and Hughes (1989, pp. 232–235). On mass markets versus highly differentiated, individualized markets, see Piore and Sabel (1984).

17. Oldenburg (1989, p. 4).
18. See, for example, Mumford (1964) and Sabel (1982).
19. Murphy (1960, p. 63).
20. Ibid., p. 115.
21. Ibid., p. 148.
22. E.g., Draper (1975).
23. Murphy and Murphy (1974).
24. For example, compare Adler and Cole (1993) with Delbridge et al. (1992), Williams et al. (1992), and Barenberg (1994, pp. 879–928).
25. The information presented on Boimondau derives primarily from Bishop (1950).
26. Ibid., p. 13.
27. Horvat (1982, pp. 172, 570).
28. For a more famous contemporary example, which continues to grow and prosper into the 1990s, see the case of the Spanish Mondragon cooperatives (Whyte and Whyte 1988). On the question of whether Mondragon's success is inescapably dependent on the unique features of Basque culture, see Whyte (1990) and Benton (1992, pp. 74–77). One possible interpretation is that Basque culture contributed significantly to the possibility of *inventing* Mondragon's institutional arrangements, but that subsequent emulation of those arrangements elsewhere need not presuppose analogous cultural traits.
29. E.g., Schrank (1978, pp. 211–243) and Horvat (1982, part IV).
30. E.g., Greenberg (1986, pp. 102–114) and Whyte and Whyte (1988, pp. 113–127).
31. Sources on Kalmar include Zimbalist (1979, pp. xx–xxi), Frampton (1980), Hill (1981, pp. 49, 104–105), and Sandberg et al. (1992, pp. 65–80). Late in 1992 Volvo suddenly announced that it was closing the Kalmar factory. The company's reasons remain contested, because the plant's productivity and output quality were better than that of Volvo's main, traditionally organized Torslanda plant (Adler and Cole 1993, p. 85; Sandberg 1993).
32. This was actually proven by Volvo's later Uddevalla plant, which adopted Kalmar's dollies within a much more democratic system of work organization (Sandberg et al. 1992, pp. 85–96).
33. Hayden (1984, pp. 189–191).
34. McCamant and Durrett (1988).
35. E.g., Alexander et al. (1977, pp. 91–334).
36. Carlson and Comstock (1986, pp. 150–231).
37. Cowan (1983), Jackson (1985), Flink (1988), and Sclove and Scheuer (1994).
38. E.g., Mansbridge (1980).
39. One of the keys to Amish stability and growth is the existence of a great

diversity of Amish and kindred Mennonite communities to which dissatisfied members can migrate without having entirely to relinquish Amish culture. The same may be true of the federated system of cooperative enterprises at Mondragon in Spain, and among Israeli kibbutzim (Whyte and Whyte 1988; Whyte and Blasi 1984).

40. E.g., Rayman (1984) and Greenberg (1986).
41. Taylor (1982). Taylor, a political scientist, probably underestimates the extent to which small-band societies have developed ritual and other means of coping with tension nonrepressively (e.g., Marshall 1976; Leacock and Lee 1982).
42. Greenberg (1986), Rothschild and Whitt (1986, pp. 169–170), and Gunn (1984, pp. 57–61, 190–192).
43. The Israeli kibbutz incorporates these qualities, but still at relatively small scale (Rosner 1983).
44. Whyte and Whyte (1988). See also Best (1990, pp. 227–240), regarding analogous institutions supporting the stability of cooperating networks of small, northern Italian manufacturing firms.
45. Whyte (1990, p. 181).
46. Bishop (1950, p. 20).
47. Mansbridge (1980, pp. 10–13).
48. See also Sirianni (1984) and Held (1991).
49. E.g., Bowles et al. (1983, chaps. 13 and 14).
50. Noble (1979, pp. 47–49); see also Piore and Sabel (1984). In fairness, Noble (1985) represents a later, pioneering probe of some of the antidemocratic consequences of military-sponsored technology development.
51. Berry et al. (1993). Nineteenth-century U.S. cities were also organized more along these lines, and popular involvement and influence in politics were correspondingly greater. Hays (1980, chap. 6) argues that subsequent, disempowering civic restructuring was an intended result of business and professional elites' reform efforts. On the general desirability of urban political decentralization, see, for example, Barber (1984, 267–273) and Bookchin (1987).
52. Oldenburg (1989), Alexander et al. (1977, pp. 70–90), and Miller (1989, pp. 202–203, 206, 356–357, 370). With regard to boundaries, some Danish cohousing communities, already solidly integrated via their well-located common house, intentionally blur their external boundaries—or situate inviting public facilities on them—in order to encourage integration with surrounding neighborhoods (McCamant and Durrett 1988, pp. 175–176, 182). By constituting natural boundaries, Switzerland's mountainous geography apparently contributed to the historic development there of communal democracy (Barber 1974, pp. 14, 105–106, 265–268). Technologies are eminently capable of surmounting such natural boundaries, but can in principle also be designed to help establish functionally analogous artificial ones.
53. E.g., Horvat (1982) and Gunn (1984, pp. 37, 97, 187–188).
54. E.g., Hirst and Zeitlin (1991, pp. 3–4, 45) and Kanter (1991).
55. E.g., Morris (1981), Hughes (1983, pp. 259–360), Piore and Sabel (1984),

Adams and Brock (1986), Chisholm (1989), and Best (1990, pp. 204, 229). Nicol (1985, pp. 195–196) predicts that modern communications technology can reduce the transaction costs that provide the classical economic rationale for corporate vertical integration.

56. Best (1990, p. 204).

57. E.g., Dutton et al. (1987).

58. Ibid. For other examples of the context-dependency of particular technologies' effects on community, see Pelto and Müller-Wille (1987) on snowmobiles and Zuboff (1988) on computerized workplaces.

59. Ihde (1979), Borgmann (1986), and Calhoun (1991).

60. See Gertler (1989, p. 279), Best (1990, pp. 235–238), and Morris (1994, pp. 219–220).

61. E.g., Mander (1978), Winn (1985), and Borgmann (1986).

62. Kegan (1982, p. 218); see also Chapter 8, on the dangers of excessively rapid social structural change.

63. There is already a disturbing precedent in the extent to which wealthy suburbanites manifest indifference to the social condition of the adjacent urban centers from which they draw material and cultural sustenance. See Bellah et al. (1985, pp. 177–181), Jackson (1985, chap. 15), and related concerns about the emerging world-managerial class of "symbolic analysts" in Reich (1992, part 3).

64. Forester (1987, pp. 69–80).

65. See Reich (1992, pp. 250–251, 268–281, 301–311).

66. Jock Gill, "Reinventing Government Processes and Political Participation," a talk given at M.I.T., on 17 February 1994. At the time, Mr. Gill was the electronic-mail coordinator in the Clinton White House.

67. Morris (1994, pp. 226–227, 232). On empowerment, see Chapter 6. With regard to transcommunity interaction, I incline toward technologies—such as video teleconferencing—in which two or more face-to-face groups communicate with one another relatively spontaneously, as opposed to systems in which participants communicate from individually isolated settings (as with today's electronic mail) or in which the conversations are managed in a nondemocratic, centralized fashion. On the other hand, Zuboff (1988, pp. 370–371) observes that computer conferencing from personal terminals can prove empowering for people who feel intimidated in face-to-face groups. That is a point worth considering. Nonetheless, interpersonal variation in face-to-face skills is at least partially class- and technology-based (Kohn 1977), and hence warrants democratic structural challenge apart from short-run tactical accommodations.

68. Mosco and Wasko (1988) and CPSR (1994).

69. Zuboff (1988, pp. 366–386).

70. This claim is supported overtly by Unger (1987) and latently—despite his overt pessimism concerning pervasive disciplinarity—by Foucault's (1984) critique of macrostructural determinism. See also Benton (1992, pp. 74–77), Best (1990, p. 161), and Whyte (1990, p. 182).

Notes to Chapter 5

1. For a nonhypothetical example, see Sandberg et al. (1992, pp. 73–74).
2. Rawls (1971, sect. 79).
3. E.g., Cowan (1983).
4. Smith (1937, p. 734).
5. Marx (1977, pp. 373–374).
6. Quoted in Zimbalist (1975, p. 51).
7. Cooley (1981) and Shaiken (1985, pp. 23–24).
8. Quoted in Green (1983, p. 295).
9. E.g., Schrank (1978, pp. 7–8, 77–83) and Delbridge et al. (1992).
10. Kohn (1977), Kohn et al. (1986), and Kohn et al. (1990).
11. E.g., Best (1990) and Sandberg et al. (1992).
12. E.g., Hirst and Zeitlin (1991) and Barenberg (1994, pp. 879–928).
13. Delbridge et al. (1992).
14. Adler and Cole (1993).
15. Sabel and Zeitlin (1985), Zuboff (1988, chap. 1), Hughes (1989, chap. 5), and Hirst and Zeitlin (1991).
16. See Chapter 10.
17. E.g., Gintis (1972, pp. 590–598) and Giddens and Held (1982). It is unquestionable that labor productivity—again, using the orthodox definition—has increased enormously during the past 2 centuries. What has not been demonstrated is that comparable increases could not also have been secured based on more convivial, self-actualizing technological means (e.g., Piore and Sabel 1984; Sabel and Zeitlin 1985).
18. Marglin (1982), Schwartz (1982), Bowles et al. (1983, chap. 6), Smith (1985, pp. 13–14), and Barenberg (1994, pp. 921–924). See also the suggestive queries in Kanter (1991, pp. 89–93).
19. See Chapter 3.
20. For the sake of argument this paragraph concedes Adler and Cole's (1993) comparative productivity measurement, although some analysts would strongly dispute or qualify it (see Williams et al. 1992; Sandberg 1993).
21. Adler and Cole (1993, pp. 91–92).
22. E.g., Uchitelle (1994).
23. E.g., Piore and Sabel (1984), Cohen and Zysman (1988), Zuboff (1988), and Sandberg et al. (1992).
24. See Cooley (1987).
25. Noble (1979); this example is about the weapons production factory discussed in Chapter 4. See also Shaiken (1985), Best (1990, pp. 156–158), and Thomas (1994).
26. Sabel (1982, pp. 70, 198–229).
27. See Chapter 4.
28. Hill (1981, p. 114).
29. E.g., Marx (1977, pp. 380, 496–497).
30. Ibid., p. 497.
31. Hill (1981).

32. Morris (1970, pp. 82–83).
33. Marx (1977, pp. 114–123).
34. See also Arendt (1958).
35. Hill (1981, pp. 97–98, 117) and Zuboff (1988, p. 271).
36. OTA (1984, pp. 185–204) and Shaiken (1985, pp. 21–22).
37. E.g., Piore and Sabel (1984) and Cooley (1987, pp. 147–154).
38. E.g., Hiatt (1990).
39. OTA (1984, chap. 4).
40. Hirst and Zeitlin (1991, pp. 5–6) and Sandberg et al. (1992, p. 287).
41. Zuboff (1988).
42. See the discussion of virtual communities in Chapter 4.
43. Cooley (1987) and Borgmann (1984).
44. Lessing (1982, pp. 13–15).
45. Yanagi (1978).
46. Within each endeavor it is not necessary that everyone have every skill, everyone participate in every decision, every decision be made and implemented by group consensus, or everyone (or, for that matter, anyone) knows everything about the entire endeavor. What matters is a broadening—rather than a totalization—of individual skills and opportunities to participate in self-management. See Sirianni (1981).
47. E.g., Kohn et al. (1990).
48. See Unger (1975). ·
49. Delbridge et al. (1992, p. 101) warn, conversely, that under circumstances of hierarchic and antagonistic labor–management relations, interchangeability could weaken workers' negotiating position. On the feasibility of sharing even senior managerial positions, see Kunde (1994).
50. On work as meditation, see Needleman (1986). On grounds for envisioning that horizontal job rotation ought generally to be an option rather than a requirement, see Sirianni (1981), Whyte and Blasi (1984, pp. 398–401), and Rothschild and Whitt (1986, pp. 109–110).
51. Rosner (1984, pp. 394–395).
52. See, for example, Sandberg et al. (1992, pp. xvi–xvii, 75–77, 231–269), Adler and Cole (1993, p. 91), Barenberg (1994, pp. 921–926), and Chapter 12.
53. Whyte and Blasi (1984, pp. 398, 401), Gunn (1984, pp. 35, 52–53, 187–188), and Rothschild and Whitt (1986, pp. 104–115). Gunn and Rothschild and Whitt assume that ideally every worker would know everything about his or her workplace. But it seems more reasonable to envision workers each being competent in one or more specific areas while otherwise being well-informed generalists. See Schutz (1946), Sirianni (1981), and Barber (1985).
54. Compare Kohn et al. (1990).
55. See the remainder of Part II.
56. E.g., compare with one another Gilligan (1982), Kegan (1982), Needleman (1982a, esp. pp. 17–22), and Lutz (1988).
57. Hausman (1979).

58. Shweder and Levine (1984, part III) and Lutz (1988).
59. E.g., Kuller (1980).
60. For hints regarding the role of emotion within moral development, see Needleman (1982b). Rudolph Steiner was one person who tried in a practical way to take technologies'—or at least architecture's—emotional influence on moral development into account (Adams 1983).
61. Weizenbaum (1976) and Turkle (1984).
62. Ihde (1979).
63. E.g., Innis (1984).
64. For hints about how sensory fragmentation might conceivably hamper moral development, see Deikman (1982).
65. Borgmann (1984).
66. Habermas (1970, chaps. 4–6).
67. Turkle (1984) describes this split but interprets it complacently. Elias (1978) locates antecedents of the split in daily rituals and utensils evolved over centuries. For spiritual–ecological, political–economic, and ecofeminist perspectives on the significance of the split, see, respectively, Duerr (1985), Hirschman (1977), and Griffin (1978).
68. On the loss of meaning with industrialism, see, for example, Kohák (1984) and Kunstler (1993).
69. E.g., Borgmann (1984, pp. 44–47).
70. See also Chapter 8. Marcuse would furthermore see this will to dominate nature as reflecting a will to dominate other people (e.g., Zimmerman 1979).

Notes to Chapter 6

1. On this specific meaning of ideology, see, for example, Geuss (1981).
2. E.g., Zimmerman (1979) and Feenberg (1991, chaps. 4, 5, and 8).
3. E.g., Habermas (1975).
4. See Chapter 2, incl. n36; and, for example, Dickson (1974), Bijker et al. (1987), and Feenberg (1991).
5. Here I follow methodologically in the footsteps of certain semioticians, cultural anthropologists, and archaeologists (e.g., Lechtman and Merrill 1977), but I depart in dwelling specifically on technologies' ideological potential.
6. E.g., Borgmann (1984).
7. This factor could conceivably recede to the extent that this sharply demarcated division of labor becomes blurred by the adoption of less hierarchic and more flexible work systems (see Best 1990).
8. Cooley (1981) and Hughes (1989, chap. 4).
9. MacKenzie (1984, p. 502).
10. Dunford (1986) and Hughes (1989, pp. 54, 139–180).

11. E.g., Nandy (1979), Headrick (1981), MacKenzie (1984, pp. 498–502), Adas (1989), and Marglin and Marglin (1990).
12. E.g., Thrupp (1989).
13. Schrank (1978, p. 226).
14. Ibid., pp. 226, 228.
15. On fatalism, see also Kohn (1977) and Kohn et al. (1990). On spillover, compare Cicourel (1981).
16. Adams and Horton (1975) and Freire (1980).
17. See Chapters 3 and 8.
18. See Chapters 11 and 12.
19. See, for example, Chapter 7.
20. For additional examples, see Edge (1973) and Ezrahi (1990, chaps. 5 and 6).
21. Harari and Bell (1982, p. xii).
22. Compare Levine (1985) and Ihde (1983, pp. 109–116).
23. Schutz (1946), Polanyi (1969), and Dreyfus and Dreyfus (1984).
24. Ihde (1979, pp. 24–25).
25. On more multidimensional electronically constituted experience, see, for example, Rheingold (1991).
26. Polanyi (1969, part III); see also Borgmann (1992, chap. 4).
27. On texts and text analogs, see Ricoeur (1977).
28. Thompson (1971, pp. 68–69).
29. E.g., Bernstein (1983), Schwartz (1984), and Barber (1985).
30. On the power relations among groups or between them and formal political institutions, see Cohen et al. (1992).
31. E.g., Cowan (1983), Wajcman (1991), and Wolf (1991).
32. Ward (1978, p. 86).
33. Nelson (1982) and Oldenburg (1989, chap. 13).
34. Foucault (1984) and Nelkin and Tancredi (1989).
35. E.g., McCluskey (1988) and Taylor (1989).
36. On the technological marginalization of poor and working-class minority groups in industrial societies—a relatively underinvestigated subject—see, for example, Billingsley (1988), Siefert et al. (1989), and Harding (1993).
37. Garrow (1986, pp. 172, 264, 692 n46).
38. E.g., Zoglin (1989).
39. E.g., Hahn (1985); Young (1989, pp. 269–271).
40. E.g., Gaventa (1980) and Angus (1988).
41. I use the prefix "neo" because it is impossible to apply Rawls's (1971) theory of distributive justice directly under situations—as here postulated—of individually legitimate, but competing and collectively insatiable, claims for redress. For instance, we are not concerned here merely with the social distribution of primary goods (governed by Rawls's second principle of justice); we also want to address the potential abridgement of fundamental human rights and liberties (governed by his even more exacting first principle).

42. E.g., Lee (1959, pp. 39–52, 89–120) and Rae et al. (1981).
43. On cities, see Jackson (1985, chap. 15), Rosen (1986), and Rose (1988). On consumption, see Elias (1978), Bourdieu (1984), and Strasser (1989).
44. Winner (1977), Chandler (1977), and Tarr (1988).
45. On alternative institutional structures, see Adams and Brock (1986), Zuboff (1988), O'Toole (1989), and Osborne and Gaebler (1993). On alternative hardware and systems, see, for example, Lovins (1977), Du Boff and Herman (1980), Piore and Sabel (1984), and Hughes (1983, chaps. 7–9).
46. Chisholm (1989).
47. Williams and Larson (1989, pp. 513–515).
48. Yates (1989).
49. Van Creveld (1989, part IV), Holstein et al. (1990), and Barnet and Cavanagh (1994).
50. E.g., OTA (1988a), Nelkin and Tancredi (1989), Mosco and Wasko (1988), and Engelberg (1993).
51. See Chapter 4.
52. Callon (1987); see also Dunford (1986), Flink (1988, chap. 19), and Hughes (1989, pp. 54, 139–183).
53. E.g., Dickson (1988, chap. 4), Wad (1988), Busch et al. (1991, chap. 6), and Greider (1993). For a contrasting, relatively sanguine assessment of TNC impacts, see Moran (1985).
54. Holstein et al. (1990); see also Barnet and Cavanagh (1994).
55. E.g., Masuda (1985).
56. See also Callon and Latour (1981) who, however, overestimate the extent to which powerful actors fully comprehend or control the durable structures they help generate.
57. Sabbah (1985, pp. 219–221) and Borgmann (1986).
58. E.g., Mosco and Wasko (1988).
59. E.g., Stefanik (1993). The converse caution—as per Criterion A—would be that communities or marginalized groups that are already solidaristic must learn, like the Old Order Amish, to resist technologies harboring appreciable nonfocal risk of promoting community disintegration (recall the story of Ibieca's water pipes; or see Sabbah 1985, pp. 219–221, and Hinton 1988).
60. Primack and von Hippel (1974, chaps. 2 and 4).
61. E.g., Hveem (1983), Krasner (1985), Botelho (1987), Kloppenburg (1988), and Mowlana (1993).
62. Compare Frug (1980) and Kierans (1983).

Notes to Chapter 7

1. Tarr (1988) and Frug (1980, esp. pp. 1065, 1139–1140).
2. The opposing view (e.g., Peterson 1982) fails to consider this possibility.
3. See Hays (1987, esp. pp. 443–445, 456–457), Porter and Brown (1991, pp.

64, 66), and Sachs (1991). When pollution thus contributes perversely to enshrining a right to further pollute, there is increased likelihood of also violating Figure 5-1's Criterion G, concerning ecological sustainability (see chap. 8).

4. See also Sagoff (1988, pp. 33–39). For further discussion of the democratic significance of externalities, which tends to enhance the grounds for Criterion E, see Chapter 10.
5. See Chapter 9.
6. E.g., Ophuls (1977).
7. Compare Arendt (1958, pp. 305, 313–320) and Lee (1959, pp. 70–77).
8. In fact, contemporary risk disputes often conceal deeper political disagreements (Casper and Wellstone 1981; Sclove 1982; Krimsky and Plough 1988).
9. E.g., Schumacher (1973), Lovins (1977), and Darrow and Saxenian (1986).
10. Brooks (1980, esp. pp. 64–66, 96).
11. Ibid., p. 67. See also Zuboff (1988) and Best (1990).
12. E.g., Adams and Brock (1986) and Best (1990). Moreover, many apparent scale economies in reality reflect large corporations' structural incentive and economic or political power to manipulate governments and markets to their own advantage (Du Boff and Herman 1980).
13. Dunford (1986, pp. 132–135).
14. See Chapter 4.
15. Brooks (1980, p. 54).
16. Hughes (1983, chap. 9).
17. See Gilman (1986), Lovins and Hubbard (1993), and also Fisk et al. (1989).
18. Gaventa (1980, chap. 8) and Frug (1980).
19. See Chapter 12.
20. Jackson (1985, pp. 146–148), Todd (1987), and Tarr (1988).
21. E.g., Kierans (1983), Castells (1985), and Greider (1993).
22. See also Broad and Cavanagh (1988, esp. pp. 99–103). By "regional," I normally mean subnational. The literature on self-reliance within developing nations tends, in contrast, to use "regional" to designate transnational sectors of the globe such as the Andean region. Some arguments pertaining to the subnational level apply with equal force transnationally, but others do not.
23. Alperovitz (1979, pp. 88–93).
24. Tickner (1987).
25. Galtung (1986).
26. D. Fischer (1985, pp. 177–178).
27. E.g., Botelho (1987), Bagchi (1988), and Whyte and Whyte (1988).
28. On transnational moral duties, see, for example, Miller et al. (1988).
29. See Chapters 3, 5, and 10.
30. Rawls (1971, sect. 65) and Agassi (1978–1979).
31. Friedman (1962, p. 109).

32. See Chapter 3 and Bowles and Gintis (1986).
33. See Borgmann (1984) and also the related discussion in Chapters 5 and 6.
34. See, for example, Hays (1969) and Frug (1980).
35. Lovins (1977, pp. 6, 58, 151–152).
36. E.g., Greenberg (1990).
37. See Harwood Group (1991) on some pertinent reasons U.S. citizens give for not more often participating in politics.
38. E.g., Hayes (1987).
39. E.g., Leckie et al. (1975).
40. See, for example (listed roughly in ascending order of their sensitivity to strong democracy), Branscomb (1993), Clinton and Gore (1993), Reich (1992), and Chapman and Yudken (1993).
41. In the interest of brevity, I will not belabor the technical practicability of greater self-reliance here. See Chapter 12, and also Coates (1981), Lovins and Lovins (1982), Todd and Todd (1984), Darrow and Saxenian (1986), and Goldhaber (1986, pp. 47–51, 135–136, 156–159).
42. E.g., Williams et al. (1987).
43. E.g., Busch et al. (1991, esp. chap. 6). The specific problem that Busch et al. highlight—the serious risk that agricultural biotechnologies could destroy the markets for developing countries' export crops—is at least as likely to arise under an international trade regime that is open as under one encompassing more locally self-reliant economies. However, for a vigorous argument against genetic engineering and other advanced biotechnologies, see Rifkin (1983).
44. See, for example, OTA (1988b), Mosco and Wasko (1988), and Abramson et al. (1988).
45. Mascuilli (1988) and Webster and Robins (1989).
46. E.g., Barber (1984, chap. 10), Abramson et al. (1988), Siefert et al. (1989, pp. 110–162), and Stefanik (1993).
47. E.g., CPSR (1994). On democratic deficiencies in the operation of modern U.S. news media, see, for example, Bagdikian (1987).
48. E.g., Barber (1984, pp. 273–278). For a critique of the specific form of electronic town meeting proposed by 1992 U.S. presidential candidate Ross Perot, see Scheuer and Sclove (1992).
49. Compare Clay (1989) and Mowlana (1993).
50. E.g., need Zuboff's (1988) envisioned democratization of the corporate panopticon remain limited to employees, when outsiders too will be affected?
51. See, for example, McChesney (1993).
52. E.g., Webster and Robins (1989) and Schiller (1993).
53. See Chapter 4.
54. Sclove and Scheuer (1994).
55. W. H. Whyte (1988, pp. 338–339, emphasis in original).
56. Arendt (1958), Alexander et al. (1977, pp. 150–186, 222–258; 297–374, 432–459), Barber (1984, pp. 267–273, 305–306), McCamant and Durrett (1988, pp. 177–182), Oldenburg (1989), and Miller (1989, pp. 200, 206, 356, 468–469, 493, 496).

57. Oldenburg (1989, pp. 71, 295–296).
58. E.g., Sorkin (1992).
59. Miller (1989, p. 206), Barber (1984, p. 268), and Oldenburg (1989).
60. See CPSR (1994).
61. Barber (1984, p. 274).
62. Fishkin (1991).

Notes to Chapter 8

1. For ambiguities and complexities in the concept of sustainability, see, for example, Norgaard (1994, esp. pp. 17–20).
2. See Chapters 4 and 7.
3. E.g., MacCormack and Strathern (1980).
4. See Chapter 3 and, on moral obligations to future generations, Kant (1970), Rawls (1971, pp. 284–298, 501–502, 512), and MacLean and Brown (1983). Sachs (1991) explains, in effect, why it is crucial that oversight be democratic.
5. See, for example, Myers (1984).
6. Some environmentalists argue that currently fashionable variants of development assistance—even sustainable development assistance—are both unjust and ecologically harmful (e.g., Sachs 1991).
7. There are, on the other hand, various arguments (e.g., from critical political theory, ecofeminism, and deep ecology) that could entail more stringent, perhaps nonanthropocentric, regimens for ecological preservation. (See, for example, Leiss 1974, Merchant 1980, and Goldsmith et al. 1988). The question is whether such arguments can be elaborated in forms that (1) more fully articulate their essential moral and practical interdependence with strong democratic norms and institutions or (2) are so morally compelling that they convincingly equal or trump democratic theory's a priori claims to moral priority (see Chapter 9).
8. See Paehlke (1990) and Hofrichter (1993).
9. See Chapter 9.
10. Kellert and Wilson (1993); see also Kohák (1984). On the other hand, Hays (1987, pp. 23–24, 36–37, 71–136, 428) suggests that the post-1960 American environmental movement was driven significantly by urban dwellers (i.e., not rural ecotopians, although perhaps their direct precursors) who had had contrasting opportunities to experience wilderness areas recreationally.
11. Kohák (1984).
12. Borgmann (1984).
13. Hays (1969, p. 272) and Sesser (1992).
14. While largely ignoring the structural role of technologies, Unger (1987) creatively weaves an entire theory of seeking more flexible social structures. However, in treating flexibility as a sufficient structural design principle, he slights other, democratically salient structural considerations such as the role of structures in constituting community, shaping moral

development, and so on. The same criticism can be made of explicit advocates of technological flexibility such as Callon (1994).

15. Marx (1977, p. 300); see also Wills (1978, pp. 123–128) on Thomas Jefferson.
16. See, for example, Lee (1959, pp. 5–58), Erikson (1976), MacIntyre (1981, pp. 98, 190–209), and Kegan (1982).
17. See, for example, Lifton (1993, esp. chap. 10).
18. Duby (1981, pp. 111–113).
19. Bijker and Aibar (n.d., pp. 548–550).
20. On ambiguities in identifying those radical innovations that are socially significant, compare Hughes (1989, esp. pp. 53–61, 181–183) with Borgmann (1984, pp. 40–44).
21. E.g., Lifton (1979).
22. Marx (1977, p. 300).
23. Compare Kant (1970, pp. 57, 85).
24. See also Kramarae (1988).
25. E.g., Delbridge et al. (1992).
26. Borgmann's (1984) "things" are promising with respect to this subcriterion because, in contrast with contemporary technological design trends, they function in a relatively self-contained, self-reliant manner. They also tend to provide multiple services and amenities simultaneously (see subcriterion [d], below).
27. See, for example, Daly (1977).
28. See Rochlin (1986).
29. See, for example, Piore and Sabel (1984).
30. See, for example, Webster and Robins (1989).
31. Hughes's (1989) culminating claim is that large, contemporary technological systems eventually suffocate further technological innovation. He presents that claim as a conclusion, but it is really only a hunch or hypothesis because his prior discussion dwells on the development of technological systems, not their subsequent structural sociotechnological effects.
32. See Hays (1969) and Tarr (1988, 176–178).
33. See Collingridge (1980).
34. Flink (1988, chap. 19).
35. Duby (1981, pp. 163–164) and Pacey (1976 pp. 25–55). For another example, recall the Dutch dike example mentioned in the preceding section.
36. For further discussion pertinent to all of this section's subcriteria, see Illich (1973), MacLean and Brown (1983), Hamlett (1992, pp. 215–218), Callon (1994), and Summerton (1994).
37. See especially Chapters 2 and 6.
38. For other arguments in favor of cultural and technological pluralism, see, for example, Illich (1973), Feyerabend (1978), Galtung (1979), Darrow and Saxenian (1986), Kramarae (1988), Banuri and Marglin (1993), and Callon (1994). For an important ethnographic study of local technological pluralism, see Chapter 12, n72.
39. On morally respectable and tolerable cultures, see Chapter 3. On

hegemonic impositions, see, for example, Casper and Wellstone (1981), Adas (1989), and Marglin and Marglin (1990).

40. For cases of relatively overt technological imperialism, see, for example, Headrick (1981), Hveem (1983), and Merchant (1989). Most such cases document resistance to the imperialism. For other examples of resistance, see Perrin (1980), Casper and Wellstone (1981), and Shiva (1991). On the possibility of assisting cultural empowerment to resist technological imperialism, recall the Berger Inquiry (chap. 3) and see also Chapters 6, 11 and 12, and generally the journal *Cultural Survival Quarterly*. On relatively uncontrolled technological transfer and social change, recall the case of Ibieca (Chapter 1) and see Hill (1988, chap. 4).

41. On respecting indigenous knowledges, technical know-how, and creativity, see, for example, Lee (1959), Rudofsky (1964), Feyerabend (1978), Janzen (1978), Gamble (1978), Schäfer (1982), Krimsky (1984), Thrupp (1989), Shiva (1991), and Banuri and Marglin (1993).

42. Nandy (1988) and Banuri and Marglin (1993). Recall Old Order Amish superiority in attending to technological polypotency (chaps. 1 and 3).

43. See, for example, Wolf (1982) and Wagner (1981).

Notes to Chapter 9

1. Sandberg et al. (1992, p. 86).
2. E.g., Lovins (1977).
3. E.g., Ford (1977).
4. See Beiner (1983), O'Cathain (1984), Dreyfus and Dreyfus (1984), and Wynne (1988).
5. E.g., Wolfe (1989, pp. 177–184, 230) and Piller (1991, pp. 189–190). On the limitations of expert-based decision making, see Chapters 3, 10, and 12.
6. See, for example, Rabinow and Sullivan (1979), Kohák (1984), and Borgmann (1984, chap. 21).
7. Sclove (1982; 1989, esp. p. 182) and de Roux (1991, pp. 43–53). On technological practice as performance, see, for example, Lechtman (1977) and Thomas (1994, chap. 7).
8. That is the aspiration, but doubtless not the accomplishment; whether it is desirable or even possible to succeed is philosophically controversial.
9. See Chapter 3.
10. E.g., Sclove (1992, pp. 144–148, including figure 2, on proposed Criteria I and J). On technology and civil liberties, see also Lovins and Lovins (1982, pp. 26–27, 160, 174, 205), OTA (1988a), and Markoff (1993). On military weapons and democracy, see, for example, Falk (1984), Noble (1985), and also Chapter 12.
11. Compare James (1958, Lectures 16 and 17). Philosophers of technology may wonder what I would say of Albert Borgmann's (1984) first-order concern that modern technological devices lead inescapably to radical

cultural impoverishment. My brief answer is that, while I greatly admire Borgmann's diagnosis (as my many citations to it attest), his key prescription—to reestablish integrative practices as the center of our lives, against a continuing background of devices—is not entirely compelling. For example, it is too individualistic and voluntaristic—too apolitical— thus downplaying the coercively disengaging structures and forces that render stable reintroduction of integrative activities into our personal lives so very difficult. Besides, even if it is possible to adopt practices that increase one's own engagement with the world or with one's family, an individual acting alone cannot reconstitute a vibrant community or polity in which to become engaged, cannot reconstitute the opportunity for challenging and creative labor in a democratic workplace, and cannot reconstitute a wilderness in which to find spiritual solace and uplift. Because strong democracy entails, for instance, a good bit of the kind of engagement that Borgmann favors (e.g., community and creative work), I see no better route to his kind of world—if indeed that is a world people wish to inhabit—than first working politically to establish the democratic structural conditions that would make that choice (as well as other substantive choices) widely practicable.

12. I am speaking of a potential objective interest; obviously some people reject strong democratic/neo-Kantian morality, and many others have never given the subject much thought.

13. E.g., Alexander et al. (1977, p. 132), Casper and Wellstone (1981), and Hofrichter (1993).

14. Weart (1988). Weart is relatively disparaging of the phenomena he describes; for a contrary interpretation, see Sclove (1989).

15. See Chapter 3, on cultural pluralism.

16. See also Sagoff (1988). For further criticism of cost–benefit analysis, see Chapter 10.

17. Kant (1959, p. 53; emphasis in original).

18. Quoted in Erikson (1976, pp. 175–176).

19. See also Polanyi (1969, pp. 123–239), Lee (1959, pp. 73–74, 164–165), and Kohák (1984).

20. See, for example, Tribe (1973) and Casper and Wellstone (1981). For a contrary instance in which sacred values and traditional discourse were fully honored, recall the Berger Inquiry (see chap. 3).

   Shrader-Frechette (1983) pragmatically defends analyzing and quantifying non-market-priced goods and bads but does not consider the moral and psychological consequences when the sacred domain is involved. She also argues (p. 154) that quantification is necessarily more compatible with democratic decisions, whereas refusal to quantify favors secretive anti-democratic decisions. I suggest, to the contrary, that there is no necessary connection. For instance, Hays (1969) relates historically how expert conservationists used technocratic analysis and management to combat popular participation in municipal democracy.

21. MacLean (1983).

Notes to Chapter 10

1. For introductions to these frameworks, see, for example, Smart and Williams (1973, pp. 3–74), Mishan (1976), and Stokey and Zeckhauser (1978). On the philosophical unity of orthodox methods of technology assessment with economism, see Tribe (1973).
2. See, for example, Gintis (1972, 1974), Schumacher (1973), Tribe (1973), Mishan (1974), Wolff (1983), Bowles and Gintis (1986), Sagoff (1988), Daly and Cobb (1989), Stirling (1993), and Norgaard (1994).
3. Rawls (1971, pp. 7, 66, 199, 259–260).
4. See Jackson (1985, chaps. 14 and 15), Wilson (1987, pp. 7–8, 100–104, 134–138), Flink (1988, chap. 19), and Davis (1992).
5. See also Hinton (1988) and Oldenburg (1989, pp. 7–8, 12, 77, 222).
6. Winner (1977, pp. 84, 127, 233–234) and Elster (1982). For other specific ways technologies influence human psychology, see Chapters 2, 5, 6, and 9.
7. Throughout this chapter, the word "economist" refers specifically to analysts whose reasoning depends on conventional neoclassical economics.
8. Stigler and Becker (1977) defend the notion of static preferences by, in effect, redefining any claimed instance of change in preference as a change in the means of satisfying a putative underlying structure of stable preferences. This begs the question of where the underlying structure comes from and avoids proving it is stable (or indeed, proving that it exists). It also introduces new layers of contestable value-laden judgments, contrary to economism's own commitment to value-neutrality (see the main text, below). For example, distinguishing a supposed stable taste from a means of satisfying it entails a value judgment that one is more important than the other (or else knowing that, psychologically, one is consistently sovereign over the other, coupled with a value judgment that behavioral consistency is all that matters in distinguishing ends from means). But does Henry love his sports car to satisfy his "stable taste" for "mobility," or is mobility now an unimportant by-product of owning a car that he has grown to experience as an extension of himself, a constitutive element of his soul? In any case, while endogenous tastes are sufficient for a decisive critique of economism, they are not necessary (e.g., my subsequent critique of economic efficiency using the concept of externalities does not depend on tastes' plasticity or endogeneity). For further discussion of the psychological effects of economic interaction and structure, see, for example, Gintis (1972, 1974), Leiss (1976), Hirsch (1978), Elster (1982), McPherson (1983), and Strasser (1989).
9. Stokey and Zeckhauser (1978, p. 4). My subsequent discussion of structuration suggests that in reality there can be no sharp separation, in theory or practice, between "working within" and "changing" a social system.
10. E.g, Gintis (1972) and Tribe (1973).
11. Strasser (1989, pp. 3–28).

12. Ibid., p. 27; see also Hughes (1989, pp. 226–243). For cautions against overestimating the influence of advertising, see Schudson (1986).

13. For the same reason, critics also fail to take into account ways in which everyday activities can transform a society's basic structure. This is true, for example, of Sagoff (1988) and of Rawls (1971), especially when Rawls insists (on pp. 8, 87–88, and 566) that his theory neither can nor needs to provide guidance regarding day-to-day social policy.

14. See Chapter 3.

15. See also Gintis (1972, 1974) and Adams and Brock (1986, pp. 29–82).

16. See the succeeding text and also Sen (1985).

17. On pure private goods, see Samuelson (1966, pp. 1223–1239) and Davis and Hulett (1977).

18. Mishan (1974, p. 75), Rosenberg (1979), and Sagoff (1988, pp. 33–40).

19. On problems raised by externalities, and on the limitations of proposed economistic methods of compensating for them, see the main text, below, and also Mishan (1974, pp. 75–86), Just et al. (1982, pp. 278–295), Sen (1985, pp. 9–13), and Sagoff (1988, pp. 33–40, 73–123). The conclusion that pervasive externalities must frustrate Pareto optimality assumes, as is reasonable, that a great many of these externalities are Pareto-relevant (Buchanan and Stubblebine 1962; Mishan 1974, pp. 75–76). In fact, determining a shadow price is necessary even in order to identify those that are not. (Hence, one cannot claim beforehand that the costs of obtaining the requisite information in a given case would be dispropor-tionate to the information's worth, because that is generally something that one could only know afterward.) Performing all the calculations needed to achieve Pareto optimality is impossible, not merely because of their great number, but also because an externality's significance varies contextually (e.g., according to the occasion, to the presence of other externalities, etc. [Sagoff 1988, p. 82]), implying that many of the subor-dinate measurement processes would be interminable (see also n31, be-low).

20. Lipsey and Lancaster (1956–1957).

21. Mishan (1981, pp. 14–21, 137).

22. This is a point not developed fully in Chapter 7, which examines only the potential, direct *political* response to externalities.

23. Cowan (1983, pp. 79–85).

24. Sclove and Scheuer (1994). In terms of modern social theory, the logic at work in these examples represents a form of collective action problem—that is, a social situation in which local or narrowly self-interested deci-sions combine to produce undesired collective results (Hardin 1982).

25. Borgmann (1984), as discussed in Chapters 8 and 9 (n11).

26. E.g., Lovins and Lovins (1982, pp. 26–27, 160, 174, 205).

27. Some such buffeting occurs independently of externalities proper; see the discussion of pecuniary externalities in, for example, Mishan (1981, pp. 134–136). Some related phenonomena are acknowledged within institu-

tional and evolutionary economics; see Callon (1994, pp. 407–409) for a succinct introduction.

28. See also Oldenburg (1989, esp. 69–70) and Strasser (1989, pp. 290–291). Compared with those discussed in Chapter 7, this paragraph refers to a broader range of structural political effects, such as democratic disempowerment that comes about independently of outside government intervention.

29. Kundera (1984, pp. 85–89).

30. Bator (1957, p. 44).

31. Mishan, who is generally astute on the subject of externalities, nevertheless overlooks their world-constitutiveness. He does so in part because, although aware that the number of externalities is unlimited, he only considers their impacts one at a time (e.g., Mishan 1976, p. 387, incl. n3). In contrast, if one looks at their combined and synthetic effects, externalities that might seem relatively unimportant in isolation from one another suddenly contribute to results of fundamental structural significance. Hence, one reason that programs to measure and internalize externalities, and so achieve Pareto optimality, must fail is that they presuppose that any given externality's significance can be assessed in isolation. Generally that is untrue, because an externality's significance is an emergent property of its dynamic, synthetic, world- and structure-constitutive relation with other externalities. Furthermore, to the extent that an externality encroaches on the sacred sphere, or exerts a structural influence on democracy, there can be no appropriate manner in which to evaluate it monetarily (see the main text, below; also see chap. 9).

32. For example, this is true—insofar as technologies function structurally—any time an economistic prescription influences technological development or deployment. See also Sagoff's (1988, pp. 74–98) discussion of how administering a willingness-to-pay study alters interviewees' prior preferences.

33. Stokey and Zeckhauser (1978, p. 257); or see the classic statement in Milton Friedman (1962, esp. chaps. 1 and 13). For moral criticism of this view, see Sagoff (1988, pp. 99–123).

34. Just et al. (1982, pp. 42–45).

35. Democratic deliberation opens the possibility of people's initial views shifting in response to argument or evidence or of creatively modifying an initial agenda in a way that helps resolve prior conflicts. Hence, there are ways to escape the dilemma of indeterminately cyclic majorities posed for economism by Arrow's (1963) Impossibility Theorem. See also Barber (1984, pp. 203–205), Mishan (1976, pp. 385–386), and Sagoff (1988, pp. 40–49, 89, 95–98).

This chapter's critique of economism may strike some as unusual in not stressing—as is true—that an economistic preoccupation with efficiency tends to slight equity. There are several reasons for not pursuing a critical strategy that, in effect, endorses the equity–efficiency construct. The most important is that the construct diverts attention from the more crucial and incisive question of establishing democratic technological and social structures. See also Sagoff (1988, pp. 57–60) and Le Grand (1990).

36. This proposed answer to economism's problems is almost suspiciously simple. Kantian theory is two centuries old; why hasn't it been elaborated this way before? A partial answer is because Kantian moral reasoning is normally applied within the domain of interpersonal ethics, not to social structures. Indeed, Marx, who did the most to develop the concept of social structures, dismissed Kant's ethics as bourgeois ideology. Meanwhile, most of those who have recently worked on critiques of economism or on non-Marxist structural theory (e.g., Anthony Giddens or Roberto Unger) are not centrally concerned with technology. Technology, however, provides a concrete point of entry for learning to develop morally informed structural design prescriptions.

37. The critique that follows will not seem entirely "external" to those enlightened welfare economists who acknowledge that a necessary adjunct to a well-ordered market economy is a democratic state (i.e., for fairly enforcing contracts, regulating monetary supply, providing public goods, administering the internalization of externalities, ensuring distributional equity, etc.). For them, trouble arises because implicitly they treat state structure as economically exogenous, in much the same way that economism also treats individual preferences as exogenous. The dilemma revealed here (both above and below) is that in reality the deep structure of the state—and of society as a whole—is, via structuration, significantly an ongoing, endogenous economic product. This basic symmetry between the dilemmas of endogenous preferences and of endogenous state/social structure arises from economism's underlying reliance on implausibly rigid, dualistic analytic distinctions and on accompanying mechanistic (as opposed to internal or dialectical) causal relations (e.g., between self and society and between economy, government, and society).

38. Compare Mishan (1974, pp. 72–76).

39. Conceivably the definition of concepts such as growth and productivity could be modified to become more directly relevant to a democratically grounded concept of social welfare. (For interesting steps in this direction, see Daly and Cobb 1989, pp. 401–455). However, to concretely specify democracy's necessary conditions involves so much contextual, qualitative judgment—including ongoing judgment in the light of unanticipated social and technological developments—that it is hard to imagine capturing it with any single quantitative indicator.

   One might think that, among economistic indicators, at least unemployment is an unequivocal social bad. By and large this is true, but a residual concern is that contemporary measures of unemployment tend to discount or deflect attention from variations in the quality of employment, including the degree to which it is democratic.

40. See also Herman (1987) and Sagoff (1988, pp. 46–48, 97).

41. This caveat relates to Sagoff's (1988, pp. 38, 195–224) regarding cost-ef-

fectiveness analysis, although he envisions a political division of labor founded on a sharper distinction between means and ends than seems practicable (structuration being one avenue by which ostensible means can in practice transform or subvert ends).

42. Individuals should be prepared, when morally or democratically necessary, to make sacrifices in their material standard of living (Rawls 1971), but not necessarily past the point where their own moral autonomy would be fundamentally degraded. The categorical imperative does not prescribe respecting one's own moral autonomy less than everyone else's. This suggests one important reason for assuring citizens an adequate minimum standard of living. Otherwise, they may reasonably balk at endorsing democratization measures that significantly risk their own moral freedom. Likewise, people currently living in extreme deprivation cannot morally be expected to place the common good before advancing their own moral condition.

43. See also Chapters 8 and 9, and Sagoff (1988).

44. E.g., Bowles et al. (1983), Kanter (1991), and Adler and Cole (1993, p. 91).

45. Logsdon (1986). It is also striking that a major ethnographic study of the Amish (Kraybill 1989) includes no index entries for "crime" or "police."

46. Stokey and Zeckhauser (1978, p. 135).

47. See also Stirling (1993).

48. It is a bit difficult to find examples of economistic works that exhibit an explicit trade-off between democracy and, say, efficiency because most simply ignore democracy altogether. Two applied policy analyses that display a trade-off are Keeny et al. (1977, pp. 312–315) and Ayres (1975). The former condones restricting certain employees' civil liberties in return for societal availability of nuclear-generated electricity. However, it begs the broader question of trading off nonemployee civil liberties by asserting that, in theory, "careful planning . . . in advance" can forestall any need for restricting civilian freedoms in response to nuclear terrorism. Ayres, in contrast, urges factoring citizens' civil liberties into cost–benefit analyses concerning nuclear power, although he judges in the case at hand that plutonium recycling should probably be prohibited rather than run a serious risk of civil liberties' abridgements. However, in fairness to these authors, they do much better than their innumerable peers who, lacking methodological encouragement or political inducement, do not even get to the stage of considering democracy as an issue.

49. E.g., Stirling (1993, pp. 98, 101). Some economistic texts, such as Just et al. (1982, pp. 278–295), observe that there are numerous difficulties associated with economism's sundry attempts to quantify externalities via such techniques as willingness-to-pay surveys. The difficulties they mention include, for example, the prohibitive cost of acquiring much of the necessary information and the problem of deterring craftily strategic responses to surveys. Even so, their list of difficulties remains seriously incomplete. For instance,

it neglects questions of distributive justice—the fact that willingness-to-pay reflects in part ability-to-pay; it fails to appreciate that the significance of a given externality tends to vary continuously as its context changes; and it is insensitive to the argument that freedom, democratic structures, and sacred values cannot be measured monetarily or treated on a par with monetized commodities.

50. For evidence, see Crane (1989, pp. 67–74).
51. Sclove (1983, pp. 41–45).
52. See also Shrader-Frechette (1985, chap. 9).
53. Suzuki (1970, p. 21).
54. See also Beiner (1983) and Schutz (1946).
55. Sagoff (1988, pp. 81–89).
56. See Chapter 3.
57. See, for example, Unger (1975) and Barber (1984, part I).
58. Sirianni (1983).
59. See also, for example, Block and Somers (1984).

Notes to Chapter 11

1. Quoted in Howard (1985, p. 46; emphasis in original).
2. See Chapter 3.
3. This point is emphasized in feminist critiques of technology, such as those of Zimmerman (1986) and Kramarae (1988). See also the discussion and notes in Chapter 6 regarding government and corporate suppression of alternatives, stifling of competition, and "elite Luddism."
4. Porter (1990, p. 75). There are several interesting historical examples of cross-fertilization in Hayden (1976, pp. 197–198, 355).
5. See also Chapter 10.
6. E.g., Kidder (1981) and Hughes (1989, pp. 75–83, 140).
7. Quoted in McCamant and Durrett (1988, p. 55).
8. Quoted in McDaniel at al. (1988, p. 7).
9. Quoted in Nelson (1982, p. 64).
10. Schrank (1978, p. 135). For further discussion of mechanized, fragmented, or rigidly supervised design processes, see Cooley (1981, 1987) and Hughes (1989, pp. 53, 181–183). Kochan (1988) reports that the "segmented, specialized, and sequential" manner in which U.S. engineering practice is organized has had adverse economic repercussions.
11. See also Hacker (1990, pp. 117, 176, 207–208).
12. Cooley (1987, pp. 119–120).
13. See Chapter 6 and also Sandberg et al. (1992, pp. 239, 262).
14. On the loss of flexibility as a technology's development or deployment continues, see Chapter 2. For evidence that a degree of flexibility may be purchased later on—albeit perhaps at high cost—see the Dutch dam example mentioned in Chapter 8.
15. E.g., Nelkin and Pollak (1979) and Casper and Wellstone (1981).

16. Schneider (1990).
17. Johnson (1993, pp. 56, 66).
18. E.g., Piller (1991) and Morris (1994, pp. 216–17).
19. Charley Richardson, of the Technology and Work Program, University of Massachusetts at Lowell (personal communication, Oct. 1993).
20. Green (1985, p. 125).
21. E.g., Bijker et al. (1987).
22. McDaniel et al. (1988, pp. 3, 9).
23. Manwaring and Wood (1984).
24. Best (1990, pp. 13–14, 138, 155–157, 164–165, 224, 235–236).
25. Collingridge (1989), Hacker (1990, chap. 6), and Ferguson (1992).
26. On Lucas, see Wainwright and Elliott (1982) and Cooley (1987). For a similar U.S. case, see Cassidy (1992, pp. 335–339).
27. Mole and Elliott (1987, chap. 3) and Mike Cooley (personal communication, Dec. 1984 and Dec. 1986).
28. Linn (1987).
29. Howard (1985), Martin (1987), and Ehn (1993, pp. 56–61).
30. Muller et al. (1992), Sandberg et al. (1992), and Schuler and Namioka (1993).
31. See Rudofsky (1964), Sommer (1983, chap. 8), Hatch (1984), and Ospina (1987).
32. See Chapter 4 and McCamant and Durrett (1988).
33. Kroll (1984).
34. See Chapter 6.
35. Hahn (1987).
36. Hingson (1982) and Kurzweil (1982).
37. Palca (1989, p. 19); see also Marx (1989) and Brown and Mikkelsen (1990).
38. Hayden (1984, pp. 63–95, 163–170).
39. Zimmerman (1986, pp. 36–37).
40. E.g., Wright (1987), Kramarae (1988), and Wajcman (1991).
41. Franklin and McNeil (1988).
42. Wharton (1980).
43. Jackson et al. (1980).
44. Spivack (1969).
45. Wolkomir (1985).
46. Nelson (1982).
47. Ward (1983, p. 119), commenting on participation in urban planning. See also Sommer (1983, pp. 119–122) and Comerio (1987).
48. See also Sandberg et al. (1992, pp. 30–37, 251–260).
49. See Chapters 3 and 10.
50. E.g., Goleman (1988) and Zuboff (1988).
51. Compare Rawls (1971), and also see Young (1989).
52. See also Walton and Gaffney (1991, pp. 108–111).
53. See also Hughes (1989, pp. 53, 180–183).
54. See also Cassidy (1992), Clement et al. (1992), and Sandberg et al. (1992).

55. Winner (1986, chap. 4) and Linn (1987).
56. See also Dickson (1988).
57. See Chapter 12 and Bok (1982, chaps. 10–14).
58. E.g, Dickson (1988, p. 221).
59. Robertson (1977) and OTA (1987, pp. 10–11).
60. See also Brooks (1993, pp. 208–214). For ongoing discussion of medical ethics review boards, see the *Hastings Center Report.*
61. Brown and Mikkelsen (1990, p. 173).
62. Thomas (1994, p. 221).
63. Instances of oppression did occur during China's Cultural Revolution (mid-1960s), although their extent is disputed. In any case, in a strong democracy protection of basic civil liberties and rights ought to preclude the sort of violent mistreatment that occurred in Maoist China.

Notes to Chapter 12

1. See Popper (1966), Landau (1969), and Sirianni (1984).
2. On methods that can be adapted to mapping, see, for example, Gaventa (1980, pp. 221–231) on oral history and Fisk et al. (1989) on inventorying latent skills and resources. For another institutional model that can potentially be adapted to debating criteria and needs, see Jennings (1990), which describes a grassroots process for developing statewide social consensus on ethical principles governing health policy.
3. Patterson (1980) and Professor David Elliott, of the Open University (personal communication, Dec. 1984). On Lucas Aerospace, see Chapter 11.
4. For information, contact NASTS, 133 Willard Building, University Park, PA 16802, U.S.A.; tel. (814) 865-3044; fax (814) 865-3047.
5. On design education, see, for example, Nelson (1982). The American Association for the Advancement of Science and the National Science Foundation both promote increased education for and employment of women and disadvantaged minorities in the sciences and engineering. That can be democratically helpful; however, it is not entirely the same as ensuring that technologies better reflect the concerns of women and minority communities, enhancing all professionals' democratic sensibilities, or increasing all students' design competencies and sensitivity to technology's political and social dimensions (Sclove 1994b, p. 17).
6. Professor John Schumacher, of Rensselaer Polytechnic Institute (personal communication, various dates, 1992–1994).
7. See Chapter 3.
8. For information on WPI's Interactive Qualifying Projects, contact The Project Center, WPI, 100 Institute Road, Worcester, MA 01609, U.S.A.; tel. (508) 831-5514; fax (508) 831-5485.
9. On collaborative research, see, for example, Whyte (1991), Fals-Borda

and Rahman (1991), Sandberg et al. (1992, esp. pp. 30–37, 100–105, 251–258), and Park et al. (1993).

10. On one technology affecting another, see the discussion of dynamic external effects in Chapter 10. On historical lessons for forecasting, see Chapter 3, n74. On latent organizational flexibility, see Chapter 6.

11. Hollander and Steneck (1990, pp. 88–89); see also Dickson (1988, pp. 227–229).

12. There are also instructive precedents and lessons for socially oriented, government-university R&D collaboration in the history of land grant colleges, state agricultural experiment stations, and cooperative extension services (e.g., Dahlberg 1986, pp. 107–162; Gerber 1992).

13. For information on U.S. STS programs, see the newsletter *Science, Technology and Society*, available from the STS Program, Maginnes Hall, Lehigh University, 9 W. Packer Ave., Bethlehem, PA 18015-3082, U.S.A.

14. E.g., Martin (1993). On the general paucity of studies of technologies' social consequences, see Rogers (1983, pp. 371–379), C. Fischer (1985), and Bolton (1990).

15. Brown and Mikkelsen (1990).

16. See Chapter 6.

17. Nelkin (1987).

18. See Step 7, below, and also Gaventa (1980, pp. 221–226), Sclove (1982, pp. 47–48), and Harwood Group (1991, pp. 13–14).

19. See, for example, Shuman and Sweig (1993), and Chapman and Yudken (1993). The Loka Institute's Technology and Democracy Project promotes a stronger grassroots, worker, and public-interest group voice in national technology decision making (including via FASTnet, the Federation of Activists on Science and Technology network). For information, contact The Loka Institute, P.O. Box 355, Amherst, MA 01004-0355, U.S.A.; e-mail loka@amherst.edu; tel. (413) 253-2828; fax (413) 253-4942.

20. See, for example, Paehlke (1990), Piller (1991, chap. 7), Lewis (1993), Cassidy and Bischak (1993, chaps. 4 and 10), and Hofrichter (1993, esp. chap. 4).

21. E.g., Herman and Chomsky (1988) and CPSR (1994).

22. Dickson (1988).

23. Sclove (1994b).

24. E.g, Lawson (1989) and Wolfe (1989, chaps. 5 and 6).

25. Schor (1992).

26. See Bowles et al. (1983, pp. 290–295, 366–370), Sirianni (1988, esp. pp. 38–45), and Gilman et al. (1993–1994).

27. On grants to nonprofessional workers, see Sirianni (1988, p. 40) and Evans et al. (1987, pp. 165–185). For additional pertinent precedents, see Schrank (1979) and *Technology for Economic Growth* (1993, p. 38).

28. See, for example, Bowles et al. (1983), Barber (1984), and Cohen et al. (1992).
29. See Chapter 6, Goldhaber (1986, pp. 130–132, 154–178), and Dutton et al. (1988, pp. 285–371).
30. Ullmann (1990, p. 31; emphasis in the original).
31. E.g., Nelkin and Pollak (1979) and Dutton et al. (1988, pp. 343–346).
32. In the sciences there is a modest precedent in programs, such as Earthwatch, that allow laypeople to participate in archaeological digs, ecological field studies, and other university research projects (Ocko 1987). This model's principal limitations are that its mode of participation is paternalistic and oriented toward affluent participants. Farm workers have sought, without much success, to influence agricultural R&D at the University of California through legal actions charging the university with allocating research monies in ways that unduly benefit agribusiness (Sun 1984; Busch et al. 1991, pp. 231–232).
33. See Comerio (1984). The government-financed Institute for Consumer Research in The Hague, Netherlands, brings together private- and public-sector engineers and designers, corporate marketing personnel, and public-interest groups to try to identify promising new products and social design criteria (Anneke Hamstra, personal communication, 11 Oct. 1994).
34. Osborne (1990).
35. Emspak (1989).
36. See Step 8, below. Within the United States, Computer Professionals for Social Responsibility has organized several major conferences about participatory design; see Muller et al. (1992) and Schuler and Namioka (1993). The federally supported Great Lakes Manufacturing Technology Center already urges companies to include workers in technology decision making (Sutherland 1993).
37. Regarding democratic deficiencies in existing mechanisms, see Dutton and Hochheimer (1982) and Comerio (1987, pp. 20, 26).
38. Dr. Frank Emspak, Visiting Fellow, NIST Advanced Technology Program (personal communication, July 1994).
39. For instance, government officials are represented in some of the quasi-public institutions that assist northern Italian manufacturers with R&D and marketing (Best 1990, p. 219).
    Jelsma and van de Poel (n.d.) outline a proposal for enlisting sympathetic scientists and engineers into preparing social assessments of their own research programs. This idea would be strengthened by encouraging researchers to enroll potentially affected people in the assessment process.
40. Bok (1982, pp. 136–152) and Brooks (1986, pp. 160–161).
41. Forester (1987, pp. 78–79). See also Rosenthal's (1990) discussion of the collective moral consequences of working under conditions of military secrecy.
42. Nelson (1989) suggests that trade secrecy is not, in practice, an important means of protecting proprietary product information. If so, it may be

economically practicable to phase out secrecy without having to establish compensatory incentives. Another approach would narrow the scope of trade secrecy protection to include only purely technical details of how a technology is produced or operates. The utility of this approach would depend partly on the extent to which such details are normally inessential to understanding a technology's external social effects (see Chapter 3). Citizen overseers could then be required to sign a limited confidentiality agreement that would permit public disclosure of any other information bearing on a technology's potential sociopolitical effects.

There are precedents for selective disclosure of information in "Right to Know" laws, which allow workers or communities to learn about their exposure to hazardous industrial materials, and in federal legislation requiring businesses with 100 or more employees to provide warning prior to a planned shutdown or large layoff (e.g., Lewis 1993). McGarity and Shapiro (1980) discuss the option of mandating public disclosure of politically salient trade secrets, coupled with legal restrictions on competitors' rights to exploit disclosed information.

An alternative remedy would weaken trade secrecy law, but increase R&D tax credits or other financial incentives even more than would be contemplated under Step 5a. However, to the extent that the effect of tax policy on R&D and innovation is not well understood, devising effective compensatory tax or financial incentives would require some policy research and experimentation.

Still another route would be to make it easier to obtain patent protection; unlike trade secrets, patent information is publicly disclosed. But a significant drawback to this route is that firms would still wish to maintain secrecy prior to submitting their patent applications (and as much as possible thereafter). Patent law would also require amendment to prevent firms from misusing patent control to block innovation (Dunford 1986; Hughes 1989, chap. 4); one means would be to make patents lapse prior to their normal 17-year term if not productively exercised.

Would relaxed trade secrecy impair international competitiveness? Probably not. For instance, even under current laws, competitors are allowed to discover secrets by reverse engineering. Thus if there were a problem, one option would be to grant limited duration tariff or patentlike protection—just long enough to cover the expected time it would otherwise have taken secrets to be found out. This would necessitate amending the General Agreement on Tariffs and Trade (see Step 11, below). Looked at from another angle, movement toward greater local self-reliance would correspondingly reduce the imperative to maintain international competitiveness (see Step 6a).

For additional ideas concerning the democratization of intellectual property rights, see Goldhaber (1986, pp. 98–99, 182–201) and Dickson (1988, pp. 330–334).

43. E.g., Dickson (1988, pp. 56–106), Anderson (1993), and Brooks (1993, p. 219).

44. See Chapters 3 and 10, Casper (1976), and Krimsky et al. (1991).
45. E.g., Krimsky (1978, p. 42), Dutton and Hochheimer (1982, pp. 13, 15), and Dickson (1988, pp. 246–258).
46. Brooks (1993, pp. 211–212); see also Step 7, below.
47. Professor Susan Cozzens, of Rensselaer Polytechnic Institute (personal communication, 15 Nov. 1993, based on research conducted collaboratively with Werner Matthiesen).
48. CCSTG (1992, p. 49).
49. Gary Chapman, of the 21st Century Project (personal communication, June 1994); see also Chapman (1994).
50. For further institutional possibilities, see Cohen et al. (1992).
51. Levine (1988).
52. See Clavel (1986), Cassidy and Bischak (1993, esp. chaps. 8 and 10), and Hofrichter (1993), as well as generally the magazines *In Context* and *The Neighborhood Works*.
53. Cassidy (1992, p. 342–343).
54. This would extend to the architectural, technological, and planning domain Barber's (1984, pp. 281–290) ideas for multichoice legislative referenda with two readings. See also McAllister (1980, pp. 241–252).
55. See, for example, Frug (1980).
56. Linn (1990); see also Coates (1981).
57. Lewis (1993, pp. 134–136). See also Hofrichter (1993) and Chapman and Yudken (1993, pp. 174–175).
58. Morris (1986) and Shavelson (1990, pp. 96–98).
59. Texas Department of Commerce (1992, pp. 2–12, 20–21) and Wisbrock (1990).
60. Ferguson and Smith, n.d.; Professor Robert J. Koester, of Ball State University (personal communication, 20 May 1993).
61. This might again entail some changes in the General Agreement on Tariffs and Trade; see Shuman (1994) and Step 11, below. The so-called new trade theory (Krugman 1990)—while almost as normatively insensitive to noneconomic structural social and political consequences of trade as the old theory—nonetheless, by challenging the sanctity of free trade ideology, helps legitimate the idea that not all instances or variants of protectionism are evil.
62. Vogel (1993, p. 14); see also Ekins (1986, pp. 194–209).
63. See, for example, Bowles et al. (1983, pp. 329–344), Shavelson (1990), and Gunn and Gunn (1991).
64. Cicourel (1981) would describe all of these activities as forms of societal "macrorepresentation." The basic strong democratic challenge is to evolve democratic procedures for generating macrorepresentations that grant primacy to structural democratic concerns.
65. See also Step 8. Alternative participatory evaluation and choice mechanisms are discussed critically in OECD (1979), McAllister (1980, pp. 235–257), Petersen (1984), Goldhaber (1986, pp. 154–217), Dutton et al. (1988), *Technology and Democracy* (n.d.), and Laird (1993a).
66. Recently a group assembled by the Ford Foundation called for a decentralized national system of Citizen Forums for cultivating civic delibera-

tion generally (Boyte and Barber 1994). Were that to come about, the activities of the proposed National Forum on Science and Technology could conceivably become one component within a more general system of Civic Forums.

67. See also Gerber (1992).

68. See, for example, Schwarz and Thompson (1990, chap. 7) and, for an interesting instance, Energy Policy Project of the Ford Foundation (1974, pp. 349–412).

69. On citizen tribunals, see, for example, Sclove (1982) and Shrader-Frechette (1985, pp. 286–315).

70. *Consensus Conference* (1992, p. 17).

71. See generally Agersnap (n.d.), Cronberg (forthcoming), Ravn (n.d.), and Vig (1992). The quotation about citizens as "final judges" is from Ravn (p. 8). The Danish Board of Technology includes 15 members chosen along the lines of the Swedish research council discussed in Step 5d (above), except that grassroots organizations are directly represented. I am indebted to Professor Norman Vig of Carleton College for helping me learn about the Danish example.

   The Board of Technology uses the consensus conference format for addressing issues of relatively narrow scope. For dealing with broader, more open-ended matters—such as alternative visions of a sustainable society—the board has learned to rely on different participatory procedures. See Andersen et al. (n.d.) and Ravn (n.d.).

72. For a variant of the model in a non-Western setting, see Janzen (1978). Janzen describes four, very different medical systems and allied epistemologies coexisting in Lower Zaire: (1) traditional magician–herbalism, (2) a form of social psychological "kinship therapy," (3) Christianity-influenced purification and initiation therapy, and (4) conventional Western medicine. In any instance of individual illness, an assemblage of concerned laypeople mediates between these competing expert traditions. Here groups of affiliated laypeople interact with one expert at a time, and each expert tradition is prevented from commanding undue deference by its coexistence with widely divergent competing traditions. Thus laypeople can take counsel from experts and yet—owing to constant far-ranging discrepancies among the experts—cannot easily evade responsibility for deliberating to reach their own collective informed judgment.

73. Even Danish consensus conferences do not reach consensus on every subissue they address (e.g., *Consensus Conference* 1992, p. 18; Cronberg, forthcoming).

74. On the latter idea, see the deliberative opinion poll mentioned at the end of Chapter 7. Small groups probably afford greater opportunity for intensive probing, deliberative depth, and mutual understanding. However, a hybrid model is also conceivable in which a large group could organize itself into subsidiary caucuses representing diverse social groups, such as women or racial and ethnic minorities. That could prove advantageous to the extent that, say, white men turned out to dominate the discussion in heterogeneous small groups (much as experts tend to dominate everyday lay citizens on advisory boards that include both). Danish consensus

conferences employ a trained meeting facilitator to combat the danger of domination within small groups (Lars Kluver, of the Danish Board of Technology, personal communication, Dec. 1994).

75. *Code of Federal Regulations*, Title 40, Sects. 1502.16 and 1508.8. I am indebted to Steve Buckley and Geoffrey Smith for bringing this regulation to my attention.

76. Federal regulations governing implementation of the National Environmental Policy Act (40 C.F.R. Part 15) contain many reasonable stipulations intended to encourage the preparation of socially useful environmental impact statements (EIS's). However, critics on both the Left and the Right argue that, in practice, EIS preparation can still degenerate into hollow proceduralism and senseless bureaucratic waste (e.g., France 1990; Piller 1991, pp. 189–190). I propose citizen oversight or participatory analysis as a means to address these concerns in the context of SPIS preparation.

77. See, for instance, Sclove and Scheuer (1994) regarding the vast range of nonusers strongly affected by the interstate highway system and likely to be affected by the proposed national information superhighway.

78. See Goldhaber (1986, pp. 195–196, 203–206).

79. See Chapter 3, regarding analogous Amish probationary trials. Government-sponsored user experiments with technological prototypes in Denmark and the Netherlands are another salient precedent (Cronberg, 1994; Schot et al., forthcoming), although they have not been designed to capture effects on nonusers.

80. See Schot (1992).

81. This assumes that local polities roughly satisfy strong democratic norms such as equality and participatory governance; when they do not, the grounds for state or federal action increase.

82. For other examples, see Schot (1992).

83. The OTA took an important step in this direction with its Science, Technology, and the Constitution project (e.g., OTA 1987). That project's principal analytic limitations were that (1) it was relatively insensitive to nonfocal effects, and (2) compatibility with the U.S. Constitution is a narrower question than overall structural compatibility with strong democracy. One of the project's strengths was that it attempted to study combined political effects of disparate technologies.

84. See, for example, OTA (1976, pp. 255–279) and, for a critical review of OTA's early history, Dickson (1988, pp. 232–243, 255–258). A more recent OTA study of adolescent health and health services took the interesting step of establishing one advisory panel composed entirely of youth aged 10 to 19 from diverse socioeconomic and experiential backgrounds (OTA 1991). However, it is difficult to tell whether the youths' views played any significant part in the study's analysis, insofar as they are summarized explicitly on just one page in a three-volume report (p. III-60).

85. There is precedent, for instance, in OTA's study of *Technology and Handicapped People* (1982). The OTA does make some effort to include representative lay stakeholders (but not other everyday citizens) in its report review process, and is even prepared to pay a modest daily honorarium, if asked. But generally experts far outweigh laypeople on OTA advisory panels and in OTA workshops; laypeople tend to find this intimidating.

86. OTA (1993, p. 71); see also van Dam and Howard (1988).

87. Rip (1988, p. 2, incl. n1).

88. I don't mean Republicans versus Democrats or pairs of teams for and against an innovation. Bipolar opposition often produces more hyperbole than insight. The important point would be to have two or more deeply divergent worldviews represented, even if all the teams should happen generally to be sympathetic (or hostile) to a particular innovation. This recommendation could obviously apply to other evaluative exercises, such as those organized by the National Research Council.

89. Dr. Philip Frankenfeld (personal communication, 2 Sept. 1993).

90. CSST (1992, 1993).

91. For early critiques, see Landau and Rosenberg (1986) and Barke (1986, pp. 89–90). For democratic critique of the Clinton initiatives, see Sclove (1994a, 1994b), Chapman and Sclove (1994), and Sclove and Scheuer (1994).

92. See, for example, OTA (1982, pp. 169–172), Dutton et al. (1988, pp. 374–379), Laird (1993a), Chapman and Yudken (1993, pp. 169–189), and Boyte and Barber (1994).

93. Dutton and Hochheimer (1982).

94. Professor Frank Laird, of the University of Denver (personal communication, 2 Sept. 1993); and Laird (1993b, pp. 26–27).

95. I think this is a fair criticism of most establishment (e.g., CCSTG 1992; Clinton and Gore 1994) and progressive (e.g., Goldhaber 1986; Chapman and Yudken 1993) strategies for guiding science and technology to address social needs.

96. See Chapter 7.

97. See Chapter 11.

98. E.g., *Technology for American Economic Growth* (1993, pp. 55–58).

99. Compare Barber (1984), Wolfe (1989), and Cohen et al. (1992).

100. In this one regard, the London Technology Networks seem to have been more consistent in supporting less paternalistic, participatory research modes. See Chapter 11.

101. Zaal and Leydesdorff (1987) and van den Broecke (1993). Also personal communication with Barend van der Meulen, of the University of Twente (Sept. and Nov. 1993); Michiel Oele, of the University of Amsterdam (Nov. 1993); and Bas L. de Boer, of the University of Amsterdam (Dec. 1993 and Oct. 1994).

    A majority of the questions addressed by science shops now involve approaches from the social sciences and humanities rather than the natural sciences or engineering. This could reflect clients' better understanding of the former disciplines or their lack of resources for following through on natural scientific or engineering research (e.g., for producing and marketing new industrial products). An alternative possibility meriting investigation is that, in terms of societal needs, mainstream corporate and government sponsors have systematically underfunded socially relevant research in the social sciences and humanities.

The Loka Institute has established a national committee to promote a U.S. network of community research centers (Sclove 1995). To contact the Loka Institute, see note 19, above.

102. Hollander (1984).
103. McLenighan (1990, pp. 36–46).
104. Walter Malakoff, of the United Brotherhood of Carpenters (personal communication, Jan. 1994).
105. Keith W. Jones, of Brookhaven National Laboratory (personal communication, Dec. 1993).
106. There are seeds of an appropriate federal funding mechanism in, for example, the Community Outreach Partnership Act of 1992.
107. This could support, incorporate, or be integrated with a decentralized National Forum on Science and Technology.
108. See, for example, Branscomb (1993, chap. 4).
109. Chapman and Yudken (1993, pp. 172–173).
110. Quoted in Cronberg (forthcoming).
111. See also Callon (1994, pp. 407–418) and Rosenberg (1994, pp. 205–207).
112. Nasar (1992).
113. On codetermination laws, see Sandberg et al. (1992). On the law and politics of democratizing corporations generally, see, for example, Hamlett (1992, chap. 11), Grossman and Adams (1993), Lewis (1993, pp. 136–141), and Barenberg (1994).
114. See the discussions of democratic macrocommunity in Chapter 3 and of organizational size limitations in Chapter 4.
115. Davenport et al. (1992, p. 53).
116. E.g., Smith (1985), van Crevald (1989), Hughes (1989, pp. 96–137, 381–442), and Markusen and Yudken (1992).
117. See, for example, Falk (1984), Noble (1985), Barus (1987), Shaw (1987), and Johansen (1993).
118. Budget figures are calculated or taken from National Science Board (1987, p. 265), Pearlstein et al. (1994), and a White House science and technology budget briefing held on 8 February 1994.
119. E.g., Keeny (1994).
120. "Defense Time Bomb" (1994); see also, for example, Schmitt (1993).
121. See Shuman and Harvey (1993, esp. chap. 4).
122. NCECD (1994).
123. Skip Stiles, legislative director, House Committee on Science, Space, and

Technology (personal communication, Aug. 1993). For further critique of the TRP, see Berkowitz (1994).

124. E.g., Smit et al. (1992).

125. See, for example, Chapman and Yudken (1993), Cassidy and Bischak (1993), and Shuman and Harvey (1993).

126. E.g., Renner (1992).

127. Picciotto (1988). For some initial effort to explore the effect of computerization and telecommunications on world financial markets, and the structural implications for democracy, see Barnet and Cavanagh (1994, pp. 359–430).

128. See also Chapter 8's suggestions concerning needed remedies for global and intercultural structural inequalities.

129. See, for example, Hveem (1983), Wad (1988), Carty (1988), and Nader et al. (1993).

130. See Nader et al. (1993) and Shuman (1994). Intriguingly, Harvard's Michael Porter (1990, p. 87) insists that—at least within developed nations—stringent product, environmental, and safety standards improve companies' abilities to compete internationally.

131. See, for example, Held (1991, esp. pp. 165–166), Makhijani (1992), and Khor (1994).

132. E.g., Stefanik (1993).

133. Kennedy (1993, pp. 51, 55).

134. Goldsmith (1993).

135. See, for example, Barnaby (1988, pp. 96–98), Sewell (1990), and Shuman and Harvey (1993, pp. 140–142).

136. See Brecher et al. (1993) and Shuman and Harvey (1993, chap. 11).

Notes to Chapter 13

1. Although this work's central purpose is prescription, some of the conceptual methods evolved here may also prove useful to scholars concerned primarily with description, explanation, or prediction. For instance, scholarly accounts of modern technology tend either to chart the development, diffusion, or influence of just one technology at a time or else lapse into heroic generalizations concerning an amorphous aggregate called "Technology." In contrast, the structurally focused design criteria developed in Part II suggest a means of integrating detailed studies of disparate technologies into comprehensive, concretely informed social interpretations of an entire technological order. (Some ethnographers attempt this when studying tribal and other small nonindustrial societies, but the aspiration hasn't been carried over to studies of more complex social formations.)

2. E.g., Bijker et al. (1987).

3. De Tocqueville (1945, vol. 1, p. 260).

4. Harwood Group (1991).

5. For indications of some of the causal relations, see Bellah et al. (1985), Kohn et al. (1986), Oldenburg (1989), and Wolfe (1989).
6. See, for example, the spring–summer 1994 issue of *Agriculture and Human Values*.
7. Brown (1993, p. B2).

BIBLIOGRAPHY

Abramson, Jeffrey B., F. Christopher Arterton, and Gary R. Orren. 1988. *The Electronic Commonwealth: The Impact of New Media Technologies on Democratic Politics*. New York: Basic Books.

Adams, David. 1983. "Architecture and Metamorphosis: The Buildings of Rudolph Steiner." *ReVision*, 6, no. 2(Fall): 48–55.

Adams, Frank, with Myles Horton. 1975. *Unearthing Seeds of Fire: The Idea of Highlander*. Winston-Salem, NC: John F. Blair.

Adams, Walter, and James W. Brock. 1986. *The Bigness Complex*. New York: Pantheon Books.

Adas, Michael. 1989. *Machines as the Measure of Men: Science, Technology, and Ideologies of Western Dominance*. Ithaca: Cornell University Press.

Adler, Paul S., and Robert E. Cole. 1993. "Designed for Learning: A Tale of Two Auto Plants." *Sloan Management Review*, 34, no. 3(Spring): 85–94.

Agassi, Judith Buber. 1978–1979. "Alienation from Work: A Conceptual Analysis." *Philosophical Forum*, 10, nos 2–4(Winter–Summer): 265–294.

Agersnap, Torben. n.d. "Consensus Conferences for Technological Assessment." In *Technology and Democracy*. Proceedings of the 3rd European Congress on Technology Assessment, Copenhagen, 4–7 November 1992. Copenhagen: TeknologiNaevnet. Vol. I, pp. 45–53.

Alcoff, Linda. 1989. "Cultural Feminism versus Post-Structuralism: The Identity Crisis in Feminist Theory." In *Feminist Theory in Practice and Process*. Eds. Micheline R. Malson et al. Chicago: University of Chicago Press. Pp. 295–326.

Alexander, Christopher, Sara Ishikawa, and Murray Silverstein, with Max Jacobson, Ingrid Fiksdahl-King, and Shlomo Angel. 1977. *A Pattern Language: Towns, Buildings, Construction*. New York: Oxford University Press.

Alperovitz, Gar. 1979. "Towards a Decentralist Commonwealth." In *Reinventing Anarchy: What Are Anarchists Thinking These Days?* Eds. Howard J. Ehrlich et al. Boston: Routledge and Kegan Paul. Pp. 84–95.

Andersen, Ida, Lise Drewes Nielsen, Morten Elle, and Olfu Danielsen. n.d. "The Scenario Workshop in Technology Assessment." In *Technology and*

Democracy. Proceedings of the 3rd European Congress on Technology Assessment, Copenhagen, 4–7 November 1992. Copenhagen: TeknologiNaevnet. Vol. II, pp. 446–455.

Anderson, Christopher. 1993. "Rocky Road for Federal Research, Inc." *Science*, 262 (22 Oct.): 496–498.

Angus, Ian H., ed. 1988. *Ethnicity in a Technological Age*. Edmonton: Canadian Institute of Ukrainian Studies, University of Alberta.

Arendt, Hannah. 1958. *The Human Condition*. Chicago: University of Chicago Press.

Arrow, Kenneth J. 1963. *Social Choice and Individual Values*. 2nd ed. New Haven, CT: Yale University Press.

Ayres, Russell W. 1975. "Policing Plutonium: The Civil Liberties Fallout." *Harvard Civil Rights–Civil Liberties Law Review*, 10(Spring): 369–443.

Bagchi, Amiya Kumar. 1988. "Technological Self-Reliance, Dependence and Underdevelopment." In *Science, Technology, and Development*. Ed. Atul Wad. Boulder, CO: Westview Press. Pp. 69–91.

Bagdikian, Ben H. 1987. *The Media Monopoly*. 2nd ed. Boston: Beacon Press.

Balabanian, Norman. 1980. "Presumed Neutrality of Technology." *Society*, 17, no. 3(March–April): 7–14.

Banuri, Tariq, and Frédérique Apffel Marglin, eds. 1993. *Who Will Save the Forests?: Knowledge, Power and Environmental Destruction*. London: Zed Books.

Barber, Benjamin. 1974. *The Death of Communal Liberty: A History of Freedom in a Swiss Mountain Canton*. Princeton, NJ: Princeton University Press.

_____. 1984. *Strong Democracy: Participatory Politics for a New Age*. Berkeley and Los Angeles: University of California Press.

_____. 1985. "The Politics of Judgment." *Raritan*, 5, no. 2(Fall): 130–143.

_____. 1992. "Jihad vs. McWorld." *Atlantic Monthly*, 269, no. 3(March): 53–65.

Barenberg, Mark. 1994. "Democracy and Domination in the Law of Workplace Cooperation: From Bureaucratic to Flexible Production." *Columbia Law Review*, 94, no. 3(April): 753–983.

Barke, Richard. 1986. *Science, Technology, and Public Policy*. Washington, DC: Congressional Quarterly Press.

Barnaby, Frank, ed. 1988. *The Gaia Peace Atlas*. New York: Doubleday.

Barnet, Richard J., and John Cavanagh. 1994. *Global Dreams: Imperial Corporations and the New World Order*. New York: Simon and Schuster.

Barus, Carl. 1987. "Military Influence on the Electrical Engineering Curriculum since World War II." *IEEE Technology and Society Magazine*, 6, no. 2(June): 3–9.

Bator, Francis M. 1957. "The Simple Analytics of Welfare Maximization." *American Economic Review*, 47, no. 1(March): 22–59.

Beiner, Ronald. 1983. *Political Judgment*. Chicago: University of Chicago Press.

Belenky, Mary Field, Blythe McVicker Clinchy, Nancy Rule Goldberger, and Jill Mattuck Tarule. 1986. *Women's Ways of Knowing: The Development of Self, Voice, and Mind*. New York: Basic Books.

Bellah, Robert N., Richard Madsen, William M. Sullivan, Ann Swidler, and Steven M. Tipton. 1985. *Habits of the Heart: Individualism and Commitment in American Life.* New York: Harper and Row.

Benton, Lauren. 1992. "The Emergence of Industrial Districts in Spain: Industrial Restructuring and Divergent Regional Responses." In *Industrial Districts and Local Economic Regeneration.* Eds. Frank Pyke and Werner Sengenberger. Geneva: International Institute for Labour Studies. Pp. 48–86.

Berger, Thomas R. 1977. *Northern Frontier, Northern Homeland: The Report of the MacKenzie Valley Pipeline Inquiry.* 2 vols. Ottawa: Minister of Supply and Services Canada.

Berkowitz, Bruce D. 1994. "Why Defense Reinvestment Won't Work." *Technology Review*, 97, no. 5(July): 53–60.

Bernard, H. Russell, and Pertti J. Pelto, eds. 1987. *Technology and Social Change.* 2nd ed. Prospect Heights, IL: Waveland Press.

Bernstein, Richard J. 1983. *Beyond Objectivism and Relativism: Science, Hermeneutics, and Praxis.* Philadelphia: University of Pennsylvania Press.

Berry, Jeffrey M., Kent E. Portney, and Ken Thompson. 1993. *The Rebirth of Urban Democracy.* Washington, DC: Brookings Institution.

Best, Michael H. 1990. *The New Competition: Institutions of Industrial Restructuring.* Cambridge, MA: Harvard University Press.

Bijker, Wiebe E., and Eduardo Aibar. n.d. "Dutch, Dikes and Democracy: An Argument against Democratic, Flexible, Good and Bad Technologies." In *Technology and Democracy.* Proceedings of the 3rd European Congress on Technology Assessment, Copenhagen, 4–7 November 1992. Copenhagen: TeknologiNaevnet. Vol. III, 538–557.

Bijker, Wiebe E., Thomas P. Hughes, and Trevor Pinch, eds. 1987. *The Social Construction of Technological Systems: New Directions in the Sociology and History of Technology.* Cambridge, MA: MIT Press.

Billingsley, Andrew. 1988. "The Impact of Technology on Afro-American Families." *Family Relations*, 37 (Oct.): 420–425.

Bishop, Claire Huchet. 1950. *All Things Common.* New York: Harper and Row.

Block, Fred, and Margaret R. Somers. 1984. "Beyond the Economistic Fallacy: The Holistic Social Science of Karl Polanyi." In *Vision and Method in Historical Sociology.* Ed. Theda Skocpol. Cambridge, UK: Cambridge University Press. Pp. 47–84.

Boehmer, Robert G., and Todd S. Palmer. 1994. "Worker Privacy in the Networked Organization: Implications of Pending U.S. and E.C. Legislation." In *The Use and Abuse of Computer Networks: Ethical, Legal, and Technological Aspects. Preliminary Report.* Washington, DC: American Association for the Advancement of Science, Sept. Pp. 143–187.

Bok, Sissela. 1982. *Secrets: On the Ethics of Concealment and Revelation.* New York: Pantheon Books.

Bolton, Roger. 1990. Review of *Technology and the Rise of the Networked City in Europe and America*, edited by Joel A. Tarr and Gabriel Dupuy. *Technology and Culture*, 31, no. 2(April): 299–301.

Bookchin, Murray. 1982. *The Ecology of Freedom: The Emergence and Dissolution of Hierarchy*. Palo Alto, CA: Cheshire Books.

_____. 1987. *The Rise of Urbanization and the Decline of Citizenship*. San Francisco: Sierra Club Books.

Booth, William. 1988. "AIDS Panels Converge on a Consensus." *Science*, 240(10 June): 1395–1396.

Borgmann, Albert. 1984. *Technology and the Character of Contemporary Life: A Philosophical Inquiry*. Chicago: University of Chicago Press.

_____. 1986. "Philosophical Reflections on the Microelectronic Revolution." In *Philosophy and Technology, Vol. II: Information Technology and Computers in Theory and Practice*. Eds. Carl Mitcham and Alois Huning. Dordrecht: D. Reidel. Pp. 189–203.

_____. 1992. *Crossing the Postmodern Divide*. Chicago: University of Chicago Press.

Botelho, Atonio José J. 1987. "Brazil's Independent Computer Strategy." *Technology Review*, 90, no. 4(May–June): 36–45.

Bourdieu, Pierre. 1984. *Distinction: A Social Critique of the Judgement of Taste*. Trans. Richard Nice. Cambridge, MA: Harvard University Press.

Bowles, Samuel, and Herbert Gintis. 1986. *Democracy and Capitalism: Property, Community, and the Contradictions of Modern Social Thought*. New York: Basic Books.

Bowles, Samuel, David M. Gordon, and Thomas E. Weisskopf. 1983. *Beyond the Wasteland: A Democratic Alternative to Economic Decline*. Garden City, NJ: Anchor Press/Doubleday.

Boyte, Harry C., and Benjamin R. Barber. 1994. "Reinventing Citizenship: The First Six Months. Report to the Ford Foundation." Minneapolis: Humphrey Institute of Public Affairs, University of Minnesota.

Branscomb, Lewis M., ed. 1993. *Empowering Technology: Implementing a U.S. Strategy*. Cambridge, MA: MIT Press.

Brecher, Jeremy, John Brown Child, and Jill Cutler, eds. 1993. *Global Visions: Beyond the New World Order*. Boston: South End Press.

Brooks, Harvey. 1980. "A Critique of the Concept of Appropriate Technology." In *Appropriate Technology and Social Values: A Critical Appraisal*. Eds. Franklin A. Long and Alexandra Oleson. Cambridge, MA: Ballinger. Pp. 53–78.

_____. 1984. "The Resolution of Technically Intensive Public Policy Disputes." *Science, Technology, and Human Values*, 9, no. 1(Winter): 39–50.

_____. 1986. "National Science Policy and Technological Innovation." In *The Positive Sum Strategy: Harnessing Technology for Economic Growth*. Eds. Ralph Landau and Nathan Rosenberg. Washington, DC: National Academy Press. Pp. 119–167.

_____. 1993. "Research Universities and the Social Contract for Science." In *Empowering Technology: Implementing a U.S. Strategy*. Ed. Lewis M. Branscomb. Cambridge, MA: MIT Press. Pp. 202–234.

Brown, George E., Jr. 1993. "Technology's Dark Side." *Chronicle of Higher Education*, 30 June, pp. B1–B2.

Brown, Phil, and Edwin J. Mikkelsen. 1990. *No Safe Place: Toxic Waste, Leukemia, and Community Action*. Berkeley and Los Angeles: University of California Press.

Brugmann, Jeb, ed. 1989. "Stalking America's International Cities." *Bulletin of Municipal Foreign Policy*, 3, no. 2(Spring): 4–6, 27–41.

Buchanan, James M., and William Craig Stubblebine. 1962. "Externality." *Economica*, 29, no. 116(Nov.): 371–384.

Busch, Lawrence, William B. Lacy, Jeffrey Burkhardt, and Laura R. Lacy. 1991. *Plants, Power, and Profit: Social, Economic, and Ethical Consequences of the New Biotechnologies*. Cambridge, MA: Blackwell.

Calhoun, Craig, with comments by Gudmund Hernes and Edward Shils. 1991. "Indirect Relationships and Imagined Communities: Large-Scale Social Integration and the Transformation of Everyday Life." In *Social Theory for a Changing Society*. Eds. Pierre Bourdieu and James S. Coleman. Boulder, CO: Westview Press. Pp. 95–130.

Callon, Michel. 1980. "Struggles and Negotiations to Define What Is Problematic and What Is Not." In *The Social Process of Scientific Investigation*. Eds. Karin D. Knorr, Roger Krohn, and Richard Whitely. *Sociology of the Sciences*, 4:197–219.

———. 1987. "Society in the Making: The Study of Technology as a Tool for Sociological Analysis." In *The Social Construction of Technological Systems*. Eds. Wiebe E. Bijker et al. Cambridge, MA: MIT Press. Pp. 83–103.

———. 1994. "Is Science a Public Good?: Fifth Mullins Lecture, Virginia Polytechnic Institute, 23 March 1993." *Science, Technology, and Human Values*, 19, no. 4(Autumn): 395–424.

Callon, Michel, and Bruno Latour. 1981. "Unscrewing the Big Leviathan: How Actors Macro-Structure Reality and How Sociologists Help Them to Do So." In *Advances in Social Theory and Methodology: Toward an Integration of Micro- and Macro-Sociologies*. Eds. K. Knorr-Cetina and A. V. Cicourel. Boston: Routledge and Kegan Paul. Pp. 277–303.

Carlson, Don, and Craig Comstock, eds. 1986. *Citizen Summitry: Keeping the Peace When It Matters Too Much to Be Left to Politicians*. Los Angeles: Jeremy P. Tarcher.

Carroll, James D. 1977. "Participatory Technology." In *Technology and Man's Future*. 2nd ed. Ed. Albert H. Teich. New York: St. Martin's Press. Pp. 336–354.

Carson, Rachel. 1962. *Silent Spring*. Boston: Houghton Mifflin.

Carty, Anthony. 1988. "Liberal Economic Rhetoric as an Obstacle to the Democratization of the World Economy." *Ethics*, 98, no. 4(July): 742–756.

Casper, Barry M. 1976. "Technology Policy and Democracy: Is the Proposed Science Court What We Need?" *Science*, 194 (1 Oct.): 29–35.

Casper, Barry M., and Paul David Wellstone. 1981. *Powerline: The First Battle of America's Energy War*. Amherst: University of Massachusetts Press.

Cassidy, Kevin J. 1992. "Defense Conversion: Economic Planning and Democratic Participation." *Science, Technology, and Human Values*, 17, no. 3(Summer): 334–348.

Cassidy, Kevin J., and Gregory A. Bischak, eds. 1993. *Real Security: Converting the Defense Economy and Building Peace.* Albany: State University of New York Press.

Cassirer, Ernst. 1963. *The Question of Jean Jacques Rousseau.* Trans. Peter Gay. Bloomington: Indiana University Press.

Castells, Manuel, ed. 1985. *High Technology, Space, and Society.* Urban Affairs Annual Reviews, vol. 28. Beverly Hills, CA: Sage.

[CCSTG]. Carnegie Commission on Science, Technology, and Government. 1992. *Enabling the Future: Linking Science and Technology to Societal Goals.* New York: Carnegie Commission on Science, Technology, and Government.

Chandler, Alfred D., Jr. 1977. *The Visible Hand: The Managerial Revolution in American Business.* Cambridge, MA: Harvard University Press.

Chapman, Gary. 1994. "The National Forum on Science and Technology Goals: Building a Democratic, Post–Cold War Science and Technology Policy." *Communications of the ACM,* 37, no. 1(Jan.): 31–37.

Chapman, Gary, and Richard Sclove. 1994. "The White House Report 'Science in the National Interest.' " Written Testimony submitted to the Subcommittee on Science of the House Committee on Science, Space, and Technology of the U.S. Congress. Austin, TX, and Amherst, MA: 21st Century Project/Loka Institute, Sept. MS, 33 pages.

Chapman, Gary, and Joel Yudken. 1993. *The 21st Century Project: Setting a New Course for Science and Technology Policy.* Palo Alto, CA: Computer Professionals for Social Responsibility.

Chisholm, Donald. 1989. *Coordination without Hierarchy: Informal Structures in Multiorganizational Systems.* Berkeley and Los Angeles: University of California Press.

Cicourel, A. V. 1981. "Notes on the Integration of Micro- and Macro-Levels of Analysis." In *Advances in Social Theory and Methodology: Toward an Integration of Micro- and Macro-Sociologies.* Eds. K. Knorr-Cetina and A. V. Cicourel. Boston: Routledge and Kegan Paul. Pp. 51–80.

[CISRD]. Committee on Industrial Support for R&D, National Science Board. 1992. *The Competitive Strength of U.S. Industrial Science and Technology: Strategic Issues.* Washington, DC: National Science Board.

Clavel, Pierre. 1986. *The Progressive City: Planning and Participation, 1969–1984.* New Brunswick, NJ: Rutgers University Press.

Clay, Jason. 1989. "Radios in the Rain Forest." *Technology Review,* 92, no. 7(Oct.): 52–57.

Clement, Andrew, Marc Griffiths, and Peter van den Besselaar. 1992. "Participatory Design Projects: A Retrospective Look." In *PDC '92. Proceedings of the Participatory Design Conference, MIT, Cambridge, Mass., 6–7 November 1992.* Eds. Michael J. Muller et al. Palo Alto, CA: Computer Professionals for Social Responsibility. Pp. 81–89.

Clinton, William J., and Albert Gore, Jr. 1993. *Technology for America's Economic Growth: A New Direction to Build Economic Strength.* Washington, DC: The White House, 22 Feb.

_____. 1994. *Science in the National Interest.* Washington, DC: Office of Science and Technology Policy, Executive Office of the President. August.

Coates, Gary J., ed. 1981. *Resettling America: Energy, Ecology and Community.* Andover, MA: Brick House.

Cobb, Roger W., and Charles D. Elder. 1972. *Participation in American Politics: The Dynamics of Agenda-Building.* Baltimore: Johns Hopkins University Press.

Cohen, Joshua, and Joel Rogers, with comments by Paul Q. Hirst et al. 1992. "Secondary Associations and Democracy." *Politics and Society,* 20, no. 4(Dec.): 389–534.

Cohen, Stephen S., and John Zysman. 1988. "Manufacturing Innovation and American Industrial Competitiveness." *Science,* 239 (4 March): 1110–1115.

Collingridge, David. 1980. *The Social Control of Technology.* New York: St. Martin's Press.

_____. 1989. "Incremental Decision Making in Technological Innovation: What Role for Science?" *Science, Technology, and Human Values,* 14, no. 2(Spring): 141–162.

Colton, Joel, and Stuart Bruchey, eds. 1987. *Technology, the Economy, and Society: The American Experience.* New York: Columbia University Press.

Comerio, Mary C. 1984. "Community Design: Idealism and Entrepreneurship." *Journal of Architectural Planning and Research,* 1, no. 4: 227–243.

_____. 1987. "Design and Empowerment: 20 Years of Community Architecture." *Built Environment,* 13, no. 1: 15–28.

Consensus Conference on Technological Animals: Final Document (preliminary issue). 1992. Copenhagen: Danish Board of Technology.

Cooley, Michael. 1981. "The Taylorisation of Intellectual Work." In *Science, Technology and the Labour Process: Marxist Studies,* vol. I. Eds. Les Levidow and Bob Young. London: CSE Books. Pp. 46–65.

_____. 1987. *Architect or Bee?: The Human Price of Technology.* Rev. ed. London: Hogarth Press.

Corn, Joseph J., ed. 1986. *Imagining Tomorrow: History, Technology, and the American Future.* Cambridge, MA: MIT Press.

Cowan, Ruth Schwartz. 1983. *More Work for Mother: The Ironies of Household Technology from the Open Hearth to the Microwave.* New York: Basic Books.

[CPSR]. Computer Professionals for Social Responsibility. 1994. "Serving the Community: A Public Interest Vision of the National Information Infrastructure." *CPSR Newsletter,* vol. 11, no. 4, and vol. 12, no. 1(Winter): 1–10, 20–23, and 26–31.

Crane, J. A. 1989. "The Problem of Valuation in Risk-Cost-Benefit Assessment of Public Policies." In *Technological Transformations: Contextual and Conceptual Implications.* Eds. Edmund F. Byrne and Joseph C. Pitts. Dordrecht: Kluwer Academic Publishers. Pp. 67–79.

Cronberg, Tarja. Forthcoming. "Technology Assessment in the Danish Socio-Political Context." *International Journal of Technology Management.*

[CSST]. Committee on Science, Space, and Technology, U.S. House of Representatives. 1992. *National Competitiveness Act of 1992: Report Together with Dissenting Views*. Report 102–841. Washington, DC: U.S. Congress.

———. 1993. *National Competitiveness Act of 1993: Report Together with Dissenting Views*. Report 103–77. Washington, DC: U.S. Congress.

Dahl, Robert A. 1985. *A Preface to Economic Democracy*. Berkeley and Los Angeles: University of California Press.

Dahl, Robert A., and Edward R. Tufte. 1973. *Size and Democracy*. Stanford: Stanford University Press.

Dahlberg, Kenneth A. 1986. *New Directions for Agriculture and Agricultural Research: Neglected Dimensions and Emerging Alternatives*. Totowa, NJ: Rowman and Allanheld.

Daly, Herman E. 1977. *Steady-State Economics: The Economics of Biophysical Equilibrium and Moral Growth*. San Francisco: W. H. Freeman.

Daly, Herman E., and John B. Cobb, Jr. 1989. *For the Common Good: Redirecting the Economy toward Community, the Environment, and a Sustainable Future*. Boston: Beacon Press.

Darrow, Ken, and Mike Saxenian. 1986. *Appropriate Technology Sourcebook: A Guide to Practical Books for Village and Small Community Technology*. Stanford, CA: Volunteers in Asia Press.

Davenport, Thomas H., Robert G. Eccles, and Laurence Prusak. 1992. "Information Politics." *Sloan Management Review*, 34, no. 1(Fall): 53–65.

Davis, J. Ronnie, and Joe R. Hulett. 1977. *Externalities, Public Goods and Mixed Goods*. Gainesville: University Presses of Florida.

Davis, Mike. 1992. "Fortress Los Angeles: The Militarization of Urban Space." In *Variations on a Theme Park: The New American City and the End of Public Space*. Ed. Michael Sorkin. Englewood Cliffs, NJ: Prentice-Hall. Pp. 154–180.

de Kadt, Martin. 1979. "Insurance: A Clerical Work Factory." In *Case Studies on the Labor Process*. Ed. Andrew Zimbalist. New York: Monthly Review Press. Pp. 242–256.

de Roux, Gustavo I. 1991. "Together Against the Computer: PAR and the Struggle of Afro-Columbians for Public Service." In *Action and Knowledge: Breaking the Monopoly with Participatory Action-Research*. Eds. Orlando Fals-Borda and Muhammad Anisur Rahman. New York: Apex Press. Pp. 37–53.

de Sola Pool, Ithiel, ed. 1983. *Forecasting the Telephone*. Norwood, NJ: Ablex.

de Tocqueville, Alexis. 1945. *Democracy in America*. 2 vols. Ed. Phillips Bradley. New York: Vintage Books.

"Defense Time Bomb." 1994. *New York Times*, 28 Jan., p. A26.

Deikman, Arthur J. 1982. *The Observing Self: Mysticism and Psychotherapy*. Boston: Beacon Press.

Delbridge, Rick, Peter Turnbull, and Barry Wilkinson. 1992. "Pushing Back the Frontiers: Management Control and Work Intensification under JIT/TQM Factory Regimes." *New Technology, Work, and Employment*, 7, no. 2(Sept.): 97–106.

Dewey, John. 1954. *The Public and Its Problems*. Chicago: Swallow Press.

Dickson, David. 1974. *Alternative Technology and the Politics of Technical Change*. Great Britain: Fontana/Collins.

———. 1988. *The New Politics of Science*. Reprint of 1984 ed., with new preface. Chicago: University of Chicago Press.

Draper, Patricia. 1975. "!Kung Women: Contrasts in Sexual Egalitarianism in Foraging and Sedentary Contexts." In *Toward an Anthropology of Women*. Ed. Rayna R. Reiter. New York: Monthly Review Press. Pp. 77–109.

Dreyfus, Hubert L., and Stuart E. Dreyfus. 1984. "From Socrates to Expert Systems: The Limits of Calculative Rationality." *Technology in Society*, 6, no. 3: 217–233.

Du Boff, Richard B., and Edward S. Herman. 1980. "Alfred Chandler's New Business History: A Review." *Politics and Society*, 10, no. 1: 87–110.

Duby, Georges. 1981. *The Age of the Cathedrals: Art and Society, 980–1420*. Trans. Eleanor Levieux and Barbara Thompson. Chicago: University of Chicago Press.

Duerr, Hans Peter. 1985. *Dreamtime: Concerning the Boundary between Wilderness and Civilization*. Trans. Felicitas Goodman. Oxford: Basil Blackwell.

Dunford, Richard. 1986. "Is the Development of Technology Helped or Hindered by Patent Law—Can Antitrust Laws Provide the Solution?" *University of New South Wales Law Journal*, 9: 117–135.

Dutton, Diana B., and John L. Hochheimer. 1982. "Institutional Biosafety Committees and Public Participation: Assessing an Experiment." *Nature*, 297(6 May): 11–15.

Dutton, Diana B., with Thomas A. Preston and Nancy E. Pfund. 1988. *Worse than the Disease: Pitfalls of Medical Progress*. Cambridge, UK: Cambridge University Press.

Dutton, William H., Jay G. Blumler, and Kenneth L. Kraemer, eds. 1987. *Wired Cities: Shaping the Future of Communications*. Boston: G. K. Hall and Co.

Edge, D. O. 1973. "Technological Metaphor." In *Meaning and Control: Essays in Social Aspects of Science and Technology*. Eds. D. O. Edge and J. N. Wolfe. London: Tavistock. Pp. 31–64.

Ehn, Pelle. 1993. "Scandinavian Design: On Participation and Skill." In *Participatory Design: Principles and Practices*. Eds. Douglas Schuler and Aki Namioka. Hillsdale, NJ: Lawrence Erlbaum Associates. Pp. 41–77.

Ekins, Paul, ed. 1986. *The Living Economy: A New Economics in the Making*. London: Routledge and Kegan Paul.

Elden, J. Maxwell. 1981. "Political Efficacy at Work: The Connection between More Autonomous Forms of Workplace Organization and a More Participatory Politics." *American Political Science Review*, 75, no. 1(March): 43–58.

Elias, Norbert. 1978. *The History of Manners*. Trans. Edmund Jephcott. New York: Pantheon Books.

Elster, John. 1982. "Sour Grapes—Utilitarianism and the Genesis of Wants." In *Utilitarianism and Beyond*. Eds. Amartya Sen and Bernard Williams. Cambridge, UK: Cambridge University Press. Pp. 219–238.

Emspak, Frank. 1989. "Flexible Automation and Skilled Labor." Paper prepared for the meeting of the International Federation for Information Processing in Prague, Czechoslovakia, June 1990. MS, 6 pages.

Energy Policy Project of the Ford Foundation. 1974. *A Time to Choose: America's Energy Future.* Cambridge, MA: Ballinger.

Engelberg, Stephen. 1993. "A New Breed of Hired Hands Cultivates Grass-Roots Anger." *New York Times,* 17 March, pp. A1, A17.

Erikson, Erik H. 1964. "The Golden Rule in Light of New Insight." In *Insight and Responsibility.* New York: W.W. Norton. Pp. 217–243.

Erikson, Kai T. 1976. *Everything in Its Path: Destruction of Community in the Buffalo Creek Flood.* New York: Simon and Schuster.

Evans, Alice Frazer, Robert A. Evans, and William Bean Kennedy. 1987. *Pedagogies for the Non-Poor.* Maryknoll, NY: Orbis Books.

Evans, Sara M., and Harry C. Boyte. 1986. *Free Spaces: The Sources of Democratic Change in America.* New York: Harper and Row.

Ezrahi, Yaron. 1990. *The Descent of Icarus: Science and the Transformation of Contemporary Democracy.* Cambridge, MA: Harvard University Press.

Falk, Richard. 1984. "Nuclear Weapons and the End of Democracy." In *Toward Nuclear Disarmament and Global Security: A Search for Alternatives.* Ed. Burns H. Weston. Boulder, CO: Westview Press. Pp. 194–204.

Fals-Borda, Orlando, and Muhammad Anisur Rahman, eds. 1991. *Action and Knowledge: Breaking the Monopoly with Participatory Action-Research.* New York: Apex Press.

Feenberg, Andrew. 1991. *Critical Theory of Technology.* Oxford: Oxford University Press.

Ferguson, David, and Leslie Smith. n.d. "Alternative Profit Potentials for Rural Indiana." Muncie, IN: Dept. of Landscape Architecture, Ball State University.

Ferguson, Eugene S. 1979. "The American-ness of American Technology." *Technology and Culture,* 20: 3–24.

————. 1992. *Engineering and the Mind's Eye.* Cambridge, MA: MIT Press.

Feyerabend, Paul K. 1978. *Science in a Free Society.* London: NLB.

Finley, M. I. 1983. *Politics in the Ancient World.* Cambridge, UK: Cambridge University Press.

Fischer, Claude S. 1985. "Studying Technology and Social Life." In *High Technology, Space, and Society.* Ed. Manuel Castells. Beverly Hills, CA: Sage. Pp. 284–300.

Fischer, Dietrich. 1985. "Defense without Threat: Switzerland's Security Policy." In *Global Militarization.* Eds. Peter Wallensteen, Johan Galtung, and Carlos Portales. Boulder, CO: Westview. Pp. 173–190.

Fishkin, James S. 1991. *Democracy and Deliberation: New Directions for Democratic Reform.* New Haven, CT: Yale University Press.

Fisk, Pliny, III, Gail Vittori, and Ray Reece. 1989. "Thinking Globally at Max's Pot." *Whole Earth Review,* no. 62(Spring): pp. 50–54.

Flink, James. 1988. *The Automobile Age.* Cambridge, MA: MIT Press.

Florman, Samuel C. 1976. *The Existential Pleasures of Engineering.* New York: St. Martin's Press.

Ford, Richard I. 1977. "The Technology of Irrigation in a New Mexico Pueblo." In *Material Culture: Styles, Organization, and Dynamics of Technology.* Eds. Heather Lechtman and Robert Merrill. New York: West. Pp. 139–154.

Forester, Tom. 1987. *High-Tech Society: The Story of the Information Technology Revolution.* Cambridge, MA: MIT Press.

Foucault, Michel. 1980. *Power/Knowledge: Selected Interviews and Other Writings, 1972–77.* Ed. Colin Gordon; Trans. Colin Gordon et al. New York: Pantheon Books.

———. 1984. *The Foucault Reader.* Ed. Paul Rabinow. New York: Pantheon Books.

Frampton, Kenneth. 1980. "The Disappearing Factory: The Volvo Experiment at Kalmar." In *Architecture for People: Explorations in a New Humane Environment.* Ed. Byron Mikellides. New York: Holt, Rinehart and Winston. Pp. 149–161.

France, Thomas. 1990. "NEPA—The Next Twenty Years." *Land and Water Law Review,* 25: 133–142.

Franklin, Sarah, and Maureen McNeil. 1988. "Reproductive Futures: Recent Literature and Current Feminist Debates on Reproductive Technologies." *Feminist Studies,* 14, no. 3(Fall): 545–560.

Freire, Paulo. 1980. *Pedagogy of the Oppressed.* Trans. Myra Bergman Ramos. New York: Continuum.

Friedman, Milton. 1962. *Capitalism and Freedom.* Chicago: University of Chicago Press.

Frug, Gerald E. 1980. "The City as a Legal Concept." *Harvard Law Review,* 93, no. 6(April): 1057–1154.

Galtung, Johan. 1979. "Towards a New International Technological Order." *Alternatives,* 4, no. 3(Jan.): 277–300.

———. 1986. "Towards a New Economics: On the Theory and Practice of Self-Reliance." In *The Living Economy: A New Economics in the Making.* Ed. Paul Ekins. London: Routledge and Kegan Paul. Pp. 97–109.

Gamble, D. J. 1978. "The Berger Inquiry: An Impact Assessment Process." *Science,* 199(3 March): 946–952.

Garrow, David J. 1986. *Bearing the Cross: Martin Luther King, Jr., and the Southern Christian Leadership Conference.* New York: William Morrow.

Gaventa, John. 1980. *Power and Powerlessness: Quiescence and Rebellion in an Appalachian Valley.* Urbana: University of Illinois Press.

Gerber, John M. 1992. "Farmer Participation in Research: A Model for Adaptive Research and Education." *American Journal of Alternative Agriculture,* 7, no. 3: 21–24.

Gertler, Len. 1989. "Telecommunications and the Changing Global Context of Urban Settlements." In *Cities in a Global Society.* Urban Affairs Annual Reviews, vol. 35. Eds. Richard V. Knight and Gary Gappert. Newbury Park, CA: Sage. Pp. 272–284.

Geuss, Raymond. 1981. *The Idea of a Critical Theory: Habermas and the Frankfurt School.* Cambridge, UK: Cambridge University Press.

Giddens, Anthony. 1979. *Central Problems in Social Theory: Action, Structure and Contradiction in Social Analysis.* Berkeley and Los Angeles: University of California Press.

Giddens, Anthony, and David Held, eds. 1982. *Classes, Power, and Conflict: Classical and Contemporary Debates.* Berkeley and Los Angeles: University of California Press.

Gilligan, Carol. 1982. *In a Different Voice: Psychological Theory and Women's Development*. Cambridge, MA: Harvard University Press.

Gilman, Robert. 1986. "Four Steps to Self-Reliance: The Story behind Rocky Mountain Institute's Economic Renewal Project." *In Context*, no. 14(Autumn): 41–46.

Gilman, Robert, et al. 1993–1994. "It's about Time." *In Context*, no. 37(Winter): 10–60.

Gintis, Herbert. 1972. "A Radical Analysis of Welfare Economics and Individual Development." *Quarterly Journal of Economics*, 86: 572–599.

_____. 1974. "Welfare Criteria with Endogenous Preferences." *International Economic Review*, 15, no. 2(June): 415–430.

Goldhaber, Michael. 1986. *Reinventing Technology: Policies for Democratic Values*. New York: Routledge and Kegan Paul.

Goldsmith, Edward. 1993. "How to Keep the Planet Habitable: The Development of the Localised and Diversified Economy." MS, 18 pages.

Goldsmith, Edward, et al. 1988. "Deep Ecology," *Ecologist*, 18, nos. 4–5: 118–185.

Goldsmith, Marc W., Ian A. Forbes, and Joe C. Turnage. 1976. *New Energy Sources: Dreams and Promises*. Framingham, MA: Energy Research Group.

Goleman, Daniel. 1988. "Why Managers Resist Machines." *New York Times*, 7 Feb., sect. 3, p. 4.

Goodwyn, Lawrence. 1978. *The Populist Moment: A Short History of the Agrarian Revolt in America*. Oxford: Oxford University Press.

Gould, Carol C. 1980. *Marx's Social Ontology: Individuality and Community in Marx's Theory of Social Reality*. Cambridge, MA: MIT Press.

_____. 1988. *Rethinking Democracy: Freedom and Social Cooperation in Politics, Economy, and Society*. Cambridge, UK: Cambridge University Press.

Gray, Paul E. 1988. "Scientific Illiteracy Threatens Democratic Process." *The Tech*, 108, no. 26(27 May): 5.

Green, Philip. 1983. "Prolegomena to a Democratic Theory of the Division of Labor." *Philosophical Forum*, 14, nos. 3–4(Spring–Summer): 263–295.

_____. 1985. *Retrieving Democracy: In Search of Civic Equality*. Totawa, NJ: Rowman and Allanheld.

Greenberg, Dolores. 1990. "Energy, Power, and Perceptions of Social Change in the Early Nineteenth Century." *American Historical Review*, 95, no. 3(June): 693–714.

Greenberg, Edward S. 1986. *Workplace Democracy: The Political Effects of Participation*. Ithaca: Cornell University Press.

Greider, William. 1993. "The Global Marketplace: A Closet Dictator." In *The Case against Free Trade*. By Ralph Nader et al. San Francisco: Earth Island Press. Pp. 195–217.

Griffin, Susan. 1978. *Woman and Nature: The Roaring Inside Her*. New York: Harper and Row.

Grossman, Richard L., and Frank T. Adams. 1993. *Taking Care of Business: Citizenship and the Charter of Incorporation*. Cambridge, MA: Charter, Ink.

Gunn, Christopher Eaton. 1984. *Workers' Self-Management in the United States*. Ithaca: Cornell University Press.

Gunn, Christopher Eaton, and Hazel Dayton Gunn. 1991. *Reclaiming Capital:*

Democratic Initiatives and Community Development. Ithaca: Cornell University Press.

Gutmann, Amy. 1987. *Democratic Education.* Princeton, NJ: Princeton University Press.

Habermas, Jürgen. 1970. *Toward a Rational Society: Student Protest, Science, and Politics.* Trans. Jeremy J. Shapiro. Boston: Beacon Press.

_____. 1975. *Legitimation Crisis.* Trans. Thomas McCarthy. Boston: Beacon Press.

Hacker, Sally L. 1990. *"Doing It the Hard Way": Investigations of Gender and Technology.* Eds. Dorothy E. Smith and Susan M. Turner. Boston: Unwin Hyman.

Hahn, Harlan. 1985. "Towards a Politics of Disability: Definitions, Disciplines, and Policies." *Social Science Journal,* 22, no. 4(Oct.): 87–105.

_____. 1987. "Civil Rights for Disabled Americans: The Foundation of a Political Agenda." In *Images of the Disabled, Disabling Images.* Eds. Alan Gartner and Tom Joe. New York: Praeger. Pp. 181–203.

Hamlett, Patrick W. 1992. *Understanding Technological Politics: A Decision-Making Approach.* Englewood Cliffs, NJ: Prentice-Hall.

Harari, Josué V. and David F. Bell. 1982. "Introduction: *Journal à plusieurs voies.*" In *Hermes: Literature, Science, Philosophy.* By Michel Serres. Baltimore: Johns Hopkins University Press. Pp. ix–xl.

Hardin, Russell. 1982. *Collective Action.* Baltimore: Johns Hopkins University Press.

Harding, Sandra, ed. 1993. *The "Racial" Economy of Science: Toward a Democratic Future.* Bloomington: Indiana University Press.

Harding, Susan Friend. 1984. *Remaking Ibieca: Rural Life in Aragon under Franco.* Chapel Hill: University of North Carolina Press.

Harwood Group. 1991. *Citizens and Politics: A View from Main Street America.* Dayton, OH: Kettering Foundation.

Hatch, C. Richard, ed. 1984. *The Scope of Social Architecture.* New York: Van Nostrand Reinhold.

Hausman, Carl R. 1979. "Criteria of Creativity." *Philosophy and Phenomenological Research,* 40, no. 2(Dec.): 237–249.

Hayden, Dolores. 1976. *Seven American Utopias: The Architecture of Communitarian Socialism, 1790–1975.* Cambridge, MA: MIT Press.

_____. 1984. *Redesigning the American Dream: The Future of Housing, Work, and Family Life.* New York: W. W. Norton.

Hayes, Peter. 1987. *Industry and Ideology: I. G. Farben in the Nazi Era.* New York: Cambridge University Press.

Hays, Samuel P. 1969. *Conservation and the Gospel of Efficiency: The Progressive Conservation Movement, 1890–1920.* Reprint of 1959 ed., with new preface. Cambridge, MA: Harvard University Press.

_____. 1980. *American Political History as Social Analysis.* Knoxville: University of Tennessee Press.

_____. 1987. *Beauty, Health, and Permanence: Environmental Politics in the United States, 1955–1985.* Cambridge, UK: Cambridge University Press.

[*Health of Research*]. *Report of the Task Force on the Health of Research.* 1992. Chairman's Report to the Committee on Science, Space, and Technology,

U.S. House of Representatives, 102nd Congress, 2nd Sess. Washington, DC: U.S. Government Printing Office, July.

Headrick, Daniel R. 1981. *The Tools of Empire: Technology and European Imperialism in the Nineteenth Century*. New York: Oxford University Press.

Held, David. 1991. "Democracy, the Nation-State and the Global System." *Economy and Society*, 20, no. 2(May): 138–172.

Herman, Edward S. 1987. "The Selling of Market Economics." In *New Ways of Knowing: The Sciences, Society, and Reconstructive Knowledge*. By Marcus G. Raskin et al. Totowa, NJ: Rowman and Littlefield. Pp. 173–199.

Herman, Edward S., and Noam Chomsky. 1988. *Manufacturing Consent: The Political Economy of the Mass Media*. New York: Pantheon Books.

Hiatt, Fred. 1990. "Custom-Made in Japan." *Washington Post National Weekly Edition*, 2–8 April, pp. 17–18.

Hill, Stephen. 1981. *Competition and Control at Work: The New Industrial Sociology*. Cambridge, MA: MIT Press.

Hill, Stephen. 1988. *The Tragedy of Technology*. London: Pluto Press.

Hingson, Michael. 1982. "The Consumer Testing Project for the Kurzweil Reading Machine for the Blind." In *Technology for Independent Living*. Proceedings of the 1980 Workshops on Science and Technology for the Handicapped. Eds. Virginia W. Stern and Martha Ross Redden. Washington, DC: American Association for the Advancement of Science. Pp. 89–90.

Hinton, Leanne. 1988. "Oral Traditions and the Advent of Electric Power." In *Technology and Women's Voices: Keeping in Touch*. Ed. Cheris Kramarae. New York: Routledge and Kegan Paul. Pp. 180–186.

Hirsch, Fred. 1978. *Social Limits to Growth*. Cambridge, MA: Harvard University Press.

Hirschman, Albert O. 1977. *The Passions and the Interests: Political Arguments for Capitalism before Its Triumph*. Princeton, NJ: Princeton University Press.

Hirst, Paul, and Jonathan Zeitlin. 1991. "Flexible Specialization versus Post-Fordism: Theory, Evidence and Policy Implications." *Economy and Society*, 20, no. 1(Feb.): 1–56.

Hochschild, Arlie Russell. 1989. *The Second Shift: Working Parents and the Revolution at Home*. New York: Viking.

Hofrichter, Richard, ed. 1993. *Toxic Struggles: The Theory and Practice of Environmental Justice*. Philadelphia: New Society Publishers.

Hollander, Rachelle. 1984. "Institutionalizing Public Service Science: Its Perils and Promise." In *Citizen Participation in Science Policy*. Ed. James C. Petersen. Amherst: University of Massachusetts Press. Pp. 75–95.

Hollander, Rachelle, and Nicholas H. Steneck. 1990. "Science-and Engineering-Related Ethics and Values Studies: Characteristics of an Emerging Field of Research." *Science, Technology, and Human Values*, 15, no. 1(Winter): 84–104.

Holstein, William J., Stanley Reed, Jonathan Kapstein, Todd Vogel, and Joseph

Weber. 1990. "The Stateless Corporation." *Business Week*, 14 May, pp. 98–106.

Holusha, John. 1994. "LTV's Weld of Worker and Manager: Steelmaker Applies a New Labor Policy." *New York Times*, 2 Aug, pp. D1, D17.

Horvat, Branko. 1982. *The Political Economy of Socialism: A Marxist Social Theory*. Armonk, NY: M. E. Sharpe.

Howard, Robert. 1985. "Utopia: Where Workers Craft New Technology." *Technology Review*, 88, no. 3(April): 43–49.

Hughes, Thomas Parke. 1983. *Networks of Power: Electrification in Western Society, 1880–1930*. Baltimore: Johns Hopkins University Press.

———. 1989. *American Genesis: A Century of Invention and Technological Enthusiasm, 1870–1970*. New York: Viking.

Hveem, Helga. 1983. "Selective Dissociation in the Technology Sector." In *Antinomies of Interdependence: National Welfare and the International Division of Labor*. Ed. John Gerard Ruggie. New York: Columbia University Press. Pp. 273–316.

Ihde, Don. 1979. *Technics and Praxis*. Boston: D. Reidel.

———. 1983. *Existential Technics*. Albany: State University of New York Press.

Illich, Ivan. 1973. *Tools for Conviviality*. New York: Harper and Row.

Innis, Robert E. 1984. "Technics and the Bias of Perception." *Philosophy and Social Criticism*, 10, no. 1(Summer): 67–89.

Jackson, Kenneth T. 1985. *Crabgrass Frontier: The Suburbanization of the United States*. New York: Oxford University Press.

[Jackson, Ted, et al.] 1980. "Users Making Choices in a Fragile Environment, Canada." In *Experiences in Appropriate Technology*. Ed. Robert J. Mitchell. Ottawa: Canadian Hunger Foundation. Pp. 47–58.

James, William. 1958. *The Varieties of Religious Experience*. New York: Mentor.

Janzen, John M. 1978. *The Quest for Therapy: Medical Pluralism in Lower Zaire*. Berkeley and Los Angeles: University of California Press.

Jelsma, Jaap, and Ibo van de Poel. n.d. "Design of Technology Assessment for Early Warning: The SESR Project at the University of Twente." In *Technology and Democracy*. Proceedings of the 3rd European Congress on Technology Assessment, Copenhagen, 4–7 November 1992. Copenhagen: TeknologiNaevnet. Vol. II, pp. 349–364.

Jennings, Bruce, ed. 1990. "Grassroots Bioethics Revisited: Health Care Priorities and Community Values." *Hastings Center Report*, 20, no. 5(Sept.–Oct.): 16–23.

Johansen, Robert C. 1993. "Military Policies and the State System as Impediments to Democracy." In *Prospects for Democracy: North, South, East, West*. Ed. David Held. Stanford, CA: Stanford University Press. Pp. 213–234.

Johnson, Peter T. 1993. "How I Turned a Critical Public into Useful Consultants." *Harvard Business Review*, Jan.–Feb, pp. 56–66.

Just, Richard E., Darrell L. Hueth, and Andrew Schmitz. 1982. *Applied Welfare Economics and Public Policy*. Englewood Cliffs, NJ: Prentice-Hall.

Kant, Immanuel. 1959. *"Foundations of the Metaphysics of Morals" and "What Is Enlightenment?"*. Trans. Lewis White Beck. Indianapolis, IN: Bobbs-Merrill.

_____. 1970. *Kant's Political Writings*. Ed. Hans Reiss; Trans. H. B. Nisbet. Cambridge, UK: Cambridge University Press.

Kanter, Rosabeth Moss, with comments by Peter Hedstrom and Edward O. Laumann. 1991. "The Future of Bureaucracy and Hierarchy in Organizational Theory: A Report from the Field." In *Social Theory for a Changing Society*. Eds. Pierre Bourdieu and James S. Coleman. Boulder, CO: Westview Press. Pp. 63–93.

Kasson, John F. 1977. *Civilizing the Machine: Technology and Republican Values in America, 1776–1900*. Harmondsworth, UK: Penguin Books.

Keeny, Spurgeon M., Jr. 1994. "Inventing an Enemy." *New York Times*, 18 June, p. 21.

Keeny, Spurgeon M., Jr., et al. 1977. *Nuclear Power Issues and Choices*. Cambridge, MA: Ballinger.

Kegan, Robert. 1982. *The Evolving Self: Problem and Process in Human Development*. Cambridge, MA: Harvard University Press.

Kellert, Stephen R., and Edward O. Wilson, eds. 1993. *The Biophilia Hypothesis*. Washington, DC, and Covelo, CA: Island Press/Shearwater Books.

Kemeny, John G. 1980. "Saving American Democracy: The Lessons of Three Mile Island." *Technology Review*, 83, no. 7(June–July): 65–75.

Kennedy, Paul. 1993. *Preparing for the Twenty-first Century*. New York: Random House.

Khor, Martin. 1994. "Operationalizing Sustainable Development in Trade." *Third World Economics*, no. 96(1–15 Sept.): 11–16.

Kidder, Tracy. 1981. *The Soul of a New Machine*. New York: Avon.

Kierans, Eric. W. 1983. "The Community and the Corporation." In *The Multinational Corporation in the 1980's*. Eds. Charles P. Kindleberger and David B. Andretsch. Cambridge, MA: MIT Press. Pp. 198–215.

Kloppenburg, Jack R., Jr., ed. 1988. *Seeds and Sovereignty: The Use and Control of Plant Genetic Resources*. Durham, NC: Duke University Press.

Kochan, Thomas A. 1988. "Adaptability of the U.S. Industrial Relations System." *Science*, 240 (15 April): 287–292.

Kohák, Erazim. 1984. *The Embers and the Stars: A Philosophical Inquiry into the Moral Sense of Nature*. Chicago: University of Chicago Press.

Kohn, Melvin L. 1977. *Class and Conformity: A Study in Values*. 2nd ed. Chicago: University of Chicago Press.

Kohn, Melvin L., Atsushi Naoi, Carrie Schoenbach, Carmi Schooler, and Kazimierz M. Slomczynski. 1990. "Position in the Class Structure and Psychological Functioning in the United States, Japan, and Poland." *American Journal of Sociology*, 95, no. 4(Jan.): 964–1008.

Kohn, Melvin L., Kazimierz M. Slomczynski, and Carrie Schoenbach. 1986. "Social Stratification and the Transmission of Values in the Family: A Cross-National Assessment." *Sociological Forum*, 1, no. 1(Winter): 73–102.

Kramarae, Cheris, ed. 1988. *Technology and Women's Voices: Keeping in Touch.* New York: Routledge and Kegan Paul.

Krasner, Stephen D. 1985. *Structural Conflict: The Third World against Global Liberalism.* Berkeley and Los Angeles: University of California Press.

Kraybill, Donald B. 1989. *The Riddle of Amish Culture.* Baltimore: Johns Hopkins University Press.

Krimsky, Sheldon. 1978. "A Citizen Court in the Recombinant DNA Debate." *Bulletin of the Atomic Scientists,* 34, no. 10(Oct.): 37–43.

———. 1984. "Epistemic Considerations on the Value of Folk-Wisdom in Science and Technology." *Policy Studies Review,* 3, no. 2(Feb.): 246–262.

Krimsky, Sheldon, James G. Ennis, and Robert Weissman. 1991. "Academic–Corporate Ties in Biotechnology: A Quantitative Study." *Science, Technology, and Human Values,* 16, no. 3(Summer): 275–287.

Krimsky, Sheldon, and Alonzo Plough. 1988. *Environmental Hazards: Communicating Risks as a Social Process.* Dover, MA: Auburn House.

Kroll, Lucien. 1984. "Anarchitecture." In *The Scope of Social Architecture.* Ed. C. Richard Hatch. New York: Van Nostrand Reinhold. Pp. 166–185.

Krugman, Paul R. 1990. *Rethinking International Trade.* Cambridge, MA: MIT Press.

Kuller, Richard. 1980. "Architecture and Emotions." In *Architecture for People: Explorations in a New Humane Environment.* Ed. Byron Mikellides. New York: Holt, Rinehart and Winston. Pp. 87–100.

Kunde, Diane. 1994. "Job-Sharing Executives Prove Two Can Do as Well as One." *Washington Post,* 23 Jan, p. H2.

Kundera, Milan. 1984. *The Unbearable Lightness of Being.* Trans. Michael Henry Heim. New York: Harper and Row.

Kunstler, James Howard. 1993. *The Geography of Nowhere: The Rise and Decline of America's Man-Made Landscape.* New York: Simon and Schuster.

Kurzweil, Raymond. 1982. "The Development of the Kurzweil Reading Machine." In *Technology for Independent Living.* Proceedings of the 1980 Workshops on Science and Technology for the Handicapped. Eds. Virginia W. Stern and Martha Ross Redden. Washington, DC: American Association for the Advancement of Science. Pp. 94–96.

Laird, Frank N. 1993a. "Participatory Analysis, Democracy, and Technological Decision Making." *Science, Technology, and Human Values,* 18, no. 3(Summer): 341–361.

———. 1993b. "The Sun Also Sets: Political Vision and the Public Construction of Solar Energy Policy." Paper prepared for delivery at the 1993 Annual Meeting of the American Political Science Association, Washington, DC, 2–5 Sept., 1993. MS, 42 pages.

Lakshmanan, Joseph L. 1990. "An Empirical Argument for Nontechnically Trained Public Members on 'Technical' Advisory Committees: FDA as a Model." *Risk: Issues in Health and Safety,* 1, no. 1(Winter): 61–74.

Landau, Martin. 1969. "Redundancy, Rationality, and the Problem of Duplication and Overlap." *Public Administration Review,* 29, no. 4(July–Aug.): 346–358.

Landau, Ralph, and Nathan Rosenberg, eds. 1986. *The Positive Sum Strategy: Harnessing Technology for Economic Growth*. Washington, DC: National Academy Press.

Lawson, Carol. 1989. "France Seen as Far Ahead in Providing Child Care." *New York Times*, 9 Nov., pp. C1, C14.

Le Grand, Julian. 1990. "Equity versus Efficiency: The Elusive Trade-Off." *Ethics*, 100, no. 3(April): 554–568.

Leacock, Eleanor, and Richard Lee, eds. 1982. *Politics and History in Band Societies*. Cambridge, UK: Cambridge University Press.

Lechtman, Heather. 1977. "Style in Technology: Some Early Thoughts." In *Material Culture: Styles, Organization and Dynamics of Technology*. Eds Heather Lechtman and Robert Merrill. New York: West. Pp. 3–20.

Lechtman, Heather, and Robert Merrill, eds. 1977. *Material Culture: Styles, Organization and Dynamics of Technology*. New York: West.

Leckie, Jim, Gil Masters, Harry Whitehouse, and Lily Young. 1975. *Other Homes and Garbage: Designs for Self-Sufficient Living*. San Francisco: Sierra Club Books.

Lee, Dorothy. 1959. *Freedom and Culture*. Englewood Cliffs, NJ: Prentice-Hall.

———. 1976. *Valuing the Self: What We Can Learn from Other Cultures*. Englewood Cliffs, NJ: Prentice-Hall.

Leiss, William. 1974. *The Domination of Nature*. Boston: Beacon Press.

———. 1976. *The Limits to Satisfaction: An Essay on the Problem of Needs and Commodities*. Toronto: University of Toronto Press.

Lessing, Doris. 1982. *The Sirian Experiments: The Report by Ambien II, of the Five*. New York: Vintage Books.

Levine, Donald N. 1985. *The Flight from Ambiguity: Essays in Social and Cultural Theory*. Chicago: University of Chicago Press.

Levine, Marc V. 1988. "Economic Development in States and Cities: Toward Democratic and Strategic Planning in State and Local Government." In *The State and Democracy: Revitalizing America's Government*. By Marc V. Levine et al. New York: Routledge. Pp. 111–146.

Lewis, Sanford J. 1993. "Ending Toxic Pollution through Environmental Democracy." In *Technology for the Common Good*. Eds Michael Shuman and Julia Sweig. Washington, DC: Institute for Policy Studies. Pp. 115–143.

Lifton, Robert Jay. 1979. *The Broken Connection: On Death and the Continuity of Life*. New York: Touchstone.

———. 1993. *The Protean Self: Human Resilience in an Age of Fragmentation*. New York: Basic Books.

Lindblom, Charles E. 1977. *Politics and Markets: The World's Political–Economic Systems*. New York: Basic Books.

Linn, Karl. 1990. "Urban Barnraising: Building Community through Environmental Restoration." *Earth Island Journal*, 5, no. 2(Spring): 34–37.

Linn, Pam. 1987. "Socially Useful Production." *Science as Culture*, no. 1, pp. 105–138.

Lipsey, R. G., and K. Lancaster. 1956–1957. "The General Theory of Second Best." *Review of Economic Studies*, 24: 11–32.

Logsdon, Gene. 1986. "Amish Economics: A Lesson for the Modern World." *Whole Earth Review*, no. 50(Spring), pp. 74–82.

Lovell, John P., and Judith Hicks Stiehm. 1989. "Military Service and Political Socialization." In *Political Learning in Adulthood: A Sourcebook of Theory and Research*. Ed. Roberta S. Sigel. Chicago: University of Chicago Press. Pp. 172–202.

Lovins, Amory B. 1977. *Soft Energy Paths: Toward a Durable Peace*. Cambridge, MA: Ballinger.

Lovins, Amory B., and Alice Hubbard. 1993 "Community Energy Planning: A Tool for Economic Development." *Environment and Development*, Feb., pp. 1–3.

Lovins, Amory B., and L. Hunter Lovins. 1982. *Brittle Power: Energy Strategy for National Security*. Andover, MA: Brick House.

Lowrance, William W. 1986. *Modern Science and Human Values*. New York: Oxford University Press.

Luria, A. R. 1976. *Cognitive Development: Its Cultural and Social Foundations*. Trans. Martin Lopez-Morillas and Lynn Solotaroff; Ed. Michael Cole. Cambridge, MA: Harvard University Press.

Lutz, Catherine A. 1988. *Unnatural Emotions: Everyday Sentiments on a Micronesian Atoll and Their Challenge to Western Theory*. Chicago: University of Chicago Press.

MacCormack, Carol P., and Marilyn Strathern, eds. 1980. *Nature, Culture and Gender*. Cambridge, UK: Cambridge University Press.

MacIntyre, Alisdair. 1981. *After Virtue: A Study in Moral Theory*. Notre Dame, IN: University of Notre Dame Press.

MacKenzie, Donald. 1984. "Marx and the Machine." *Technology and Culture*, 25, no. 3(July): 473–502.

MacLean, Douglas. 1983. "Valuing Human Life." In *Uncertain Power: The Struggle for a National Energy Policy*. Ed. Dorothy S. Zinberg. New York: Pergamon Press. Pp. 93–111.

MacLean, Douglas, and Peter G. Brown, eds. 1983. *Energy and the Future*. Totowa, NJ: Rowman and Littlefield.

Makhijani, Arjun. 1992. *From Global Capitalism to Economic Justice*. New York: Apex Press.

Mander, Jerry. 1978. *Four Arguments for the Elimination of Television*. New York: Morrow Quill.

Mansbridge, Jane J. 1980. *Beyond Adversary Democracy*. New York: Basic Books.

Manwaring, Tony, and Stephen Wood. 1984. "The Ghost in the Machine: Tacit Skills in the Labor Process." *Socialist Review*, 14, no. 2(March–April): 57–83, 94.

Marglin, Frédérique Apffel, and Stephen A. Marglin, eds. 1990. *Dominating Knowledge: Development, Culture, and Resistance*. Oxford: Clarendon Press.

Marglin, Stephen A. 1982. "What Do Bosses Do? The Origins and Functions of Hierarchy in Capitalist Production." In *Classes, Power, and Conflict: Classical and Contemporary Debates*. Eds. Anthony Giddens and David Held. Berkeley and Los Angeles: University of California Press. Pp. 285–298.

Markoff, John. 1993. "U.S. as Big Brother of Computer Age." *New York Times*, 6 May, pp. D1, D7.

Markusen, Ann, and Joel Yudken. 1992. *Dismantling the Cold War Economy*. New York: Basic Books.

Marshall, Lorna. 1976. "Sharing, Talking, and Giving: Relief of Social Tensions among the !Kung." In *Kalahari Hunter–Gatherers: Studies of the !Kung San and Their Neighbors*. Eds. Richard B. Lee and Irven DeVore. Cambridge, MA: Harvard University Press. Pp. 349–371.

Martin, Andrew. 1987. "Unions, the Quality of Work, and Technological Change in Sweden." In *Worker Participation and the Politics of Reform*. Ed. Carmen J. Sirianni. Philadelphia: Temple University Press. Pp. 95–139.

Martin, Brian. 1993. "The Critique of Science Becomes Academic." *Science, Technology, and Human Values*, 18, no. 2(Spring): 247–259.

Marx, Jean L. 1989. "The Trials of Conducting AIDS Drugs Trials." *Science*, 244(26 May): 916–918.

Marx, Karl. 1977. *Karl Marx: Selected Writings*. Ed. David MacLellan. Oxford: Oxford University Press.

Masciulli, Joseph. 1988. "Rousseau versus Instant Government: Democratic Participation in the Age of Telepolitics." In *Democratic Theory and Technological Society*. Eds. Richard R. Day, Ronald Beiner, and Joseph Masciulli. New York: M. E. Sharpe. Pp. 150–164.

Masuda, Yoneji. 1985. "Computopia." In *The Information Technology Revolution*. Ed. Tom Forester. Cambridge, MA: MIT Press. Pp. 620–634.

Mazur, Allan. 1986. "Controlling Technology." In *Technology and Man's Future*. Ed. Albert H. Teich. 4th ed. New York: St. Martin's Press. Pp. 245–258.

McAllister, Donald M. 1980. *Evaluation of Environmental Planning: Assessing Environmental, Social, Economic, and Political Trade-Offs*. Cambridge, MA: MIT Press.

McCamant, Kathryn, and Charles Durrett. 1988. *Cohousing: A Contemporary Approach to Housing Ourselves*. Berkeley, CA: Habitat Press.

McChesney, Robert W. 1993. *Telecommunications, Mass Media, and Democracy: The Battle for the Control of U.S. Broadcasting, 1928–1935*. New York: Oxford University Press.

McCluskey, Martha T. 1988. "Rethinking Equality and Difference: Disability Discrimination in Public Transportation." *Yale Law Journal*, 97, no. 5(April): 863–880.

McDaniel, Susan A., Helen Cummins, and Rachelle Sender Beauchamp. 1988. "Mothers of Invention?: Meshing the Roles of Inventor, Mother, and Worker." *Women's Studies International Forum*, 11, no. 1: 1–12.

McGarity, Thomas O., and Sidney A. Shapiro. 1980. "The Trade Secret Status of Health and Safety Testing Information: Reforming Agency Disclosure Policies." *Harvard Law Review*, 93, no. 5(March): 837–88.

McLenighan, Valjean. 1990. *Sustainable Manufacturing: Saving Jobs, Saving the Environment*. Chicago: Center for Neighborhood Technology.

McLuhan, Marshall. 1964. *Understanding Media: The Extensions of Man*. New York: Mentor.

McPherson, Michael. 1983. "Want Formation, Morality, and Some 'Interpretive' Aspects of Economic Inquiry." In *Social Science as Moral Inquiry*. Eds. Norma Haan et al. New York: Columbia University Press. Pp. 96–124.

Merchant, Carolyn. 1980. *The Death of Nature: Women, Ecology, and the Scientific Revolution*. San Francisco: Harper and Row.

———. 1989. *Ecological Revolutions: Nature, Gender, and Science in New England*. Chapel Hill: University of North Carolina Press.

Mill, John Stuart. 1972. *"Utilitarianism," "On Liberty," and "Considerations on Representative Government"*. Ed. H. B. Acton. New York: E. P. Dutton.

Miller, David, et al. 1988. "Symposium on Duties beyond Borders." *Ethics*, 98, no. 4(July): 647–756.

Miller, Donald L. 1989. *Lewis Mumford: A Life*. New York: Weidenfeld and Nicolson.

Mishan, E. J. 1974. "Ills, Bads, and Disamenities: The Wages of Growth." *Daedalus*, 102, no. 4(Fall): 63–87.

———. 1976. *Cost–Benefit Analysis*. 2nd ed. New York: Praeger.

———. 1981. *Economic Efficiency and Social Welfare: Selected Essays on Fundamental Aspects of the Economic Theory of Social Welfare*. London: George Allen and Unwin.

Mole, Veronica, and Dave Elliott. 1987. *Enterprising Innovation: An Alternative Approach*. London: Frances Pinter.

Moran, Theodore H., ed. 1985. *Multinational Corporations: The Political Economy of Foreign Direct Investment*. Lexington, MA: Lexington Books.

Morgan, Arthur E. 1984. *The Small Community: Foundation of Democratic Life*. Yellow Springs, OH: Community Service.

Morris, David. 1981. "Self-Reliant Cities: The Rise of the New City-States." In *Resettling America: Energy, Ecology, and the Community*. Ed. Gary J. Coates. Andover, MA: Brick House. Pp. 240–262.

———. 1986. "The Self-Reliant City: St. Paul, MN." *Changing Work*, Spring, pp. 8–11.

———. 1994. "Communities: Building Authority, Responsibility, and Capacity." In *State of the Union 1994: The Clinton Administration and the Nation in Profile*. Eds. Richard Caplan and John Feffer. Boulder, CO: Westview Press. Pp. 214–233.

Morris, William. 1970. *News from Nowhere; or, An Epoch of Rest*. London: Routledge and Kegan Paul.

Mosco, Vincent, and Janet Wasko, eds. 1988. *The Political Economy of Information*. Madison: University of Wisconsin Press.

Mowlana, Hamid. 1993. "Toward a NWICO for the Twenty-first Century?" *Journal of International Affairs*, 47, no. 1(Summer): 59–72.

Muller, Michael J., Sarah Kuhn, and Judith A. Meskill, eds. 1992. *PDC '92*. Proceedings of the Participatory Design Conference, MIT, Cambridge, Mass., 6–7 Nov. 1992. Palo Alto, CA: Computer Professionals for Social Responsibility.

Mumford, Lewis. 1961. *The City in History: Its Origins, Its Transformations, and Its Prospects.* New York: Harcourt, Brace and World.

———. 1964. "Authoritarian and Democratic Technics." *Technology and Culture,* 5, no. 1(Winter): 1–8.

———. 1967. *Technics and Human Development: The Myth of the Machine.* Vol. I. New York: Harcourt, Brace and World.

Murphy, Robert F. 1960. *Headhunter's Heritage: Social and Economic Change among the Mundurucú Indians.* Berkeley and Los Angeles: University of California Press.

Murphy, Yolanda, and Robert F. Murphy. 1974. *Women of the Forest.* New York: Columbia University Press.

Myers, Norman, ed. 1984. *Gaia: An Atlas of Planet Management.* Garden City, NY: Anchor Press/Doubleday.

Nader, Ralph, et al. 1993. *The Case against Free Trade.* San Francisco: Earth Island Press.

Nandy, Ashis. 1979. "The Traditions of Technology." *Alternatives,* 4, no. 3(Jan.): 371–385.

———, ed. 1988. *Science, Hegemony and Violence: A Requiem for Modernity.* Delhi: Oxford University Press

Nasar, Sylvia. 1992. "Even among the Well-Off, the Richest Get Richer." *New York Times,* 5 March, pp. A1, D24.

National Science Board. 1987. *Science and Engineering Indicators—1987.* NSB 87–1. Washington, DC: U.S. Government Printing Office.

[NCECD]. National Commission for Economic Conversion and Disarmament. 1994. *Conversion Communications Forum,* 1, no. 1(April). Washington, DC: NCECD.

Needleman, Carla. 1986. *The Work of Craft: An Inquiry in the Nature of Crafts and Craftsmanship.* London: Arkana.

Needleman, Jacob. 1982a. *Consciousness and Tradition.* New York: Crossroad.

———. 1982b. *The Heart of Philosophy.* New York: Alfred A. Knopf.

Nelkin, Dorothy. 1987. *Science, Technology and the Press.* New York: W. H. Freeman.

———, ed. 1992. *Controversy: Politics of Technical Decisions.* 3rd ed. Newbury Park, CA: Sage.

Nelkin, Dorothy, and Michael Pollak. 1979. "Public Participation in Technological Decisions: Reality or Grand Illusion." *Technology Review,* 81, no. 8(Aug.–Sept.): 55–64.

Nelkin, Dorothy, and Laurence Tancredi. 1989. *Dangerous Diagnostics: The Social Power of Biological Information.* New York: Basic Books.

Nelson, Doreen. 1982. *City Building Education: A Way to Learn.* Santa Monica, CA: Center for City Building Education Programs.

Nelson, Richard R. 1989. "What Is Private and What Is Public about Technology?" *Science, Technology and Human Values,* 14, no. 3(Summer): 229–241.

Nicol, Lionel. 1985. "Communications Technology: Economic and Spatial Impacts." In *High Technology, Space, and Society.* Ed. Manuel Castells. Beverly Hills, CA: Sage. Pp. 191–209.

Noble, David F. 1979. "Social Choice in Machine Design: The Case of Automatically Controlled Machine Tools." In *Case Studies on the Labor*

Process. Ed. Andrew Zimbalist. New York: Monthly Review Press. Pp. 18–50.

————. 1985. "Command Performance: A Perspective on Military Enterprise and Technological Change." In *Military Enterprise and Technological Change.* Ed. Merritt Roc Smith. Cambridge, MA: MIT Press. Pp. 329–345.

Norgaard, Richard B. 1994. *Development Betrayed: The End of Progress and a Coevolutionary Revisioning of the Future.* London: Routledge.

Nussbaum, Martha, et al. 1994. "Patriotism or Cosmopolitanism?" *Boston Review,* 19, no. 5(Oct.–Nov.): 3–34.

O'Cathain, Conall S. 1984. "Why Is Design Logically Impossible?" In *Design Theory and Practice.* Eds. Richard Langdon and Patrick A. Purcell. London: Design Council. Pp. 33–36.

O'Toole, Laurence J., Jr. 1989. "Alternative Mechanisms for Multiorganizational Implementation: The Case of Wastewater Management." *Administration and Society,* 21, no. 3(Nov.): 313–339.

Ocko, Stephanie. 1987. "The Business of Selling Adventure." *Technology Review,* 90, no. 2(Feb.–March): 64–73.

[OECD]. Organisation for Economic Cooperation and Development. 1979. *Technology on Trial: Public Participation in Decision-Making Related to Science and Technology.* Paris: OECD.

Oldenburg, Ray. 1989. *The Great Good Place: Cafés, Coffee Shops, Community Centers, Beauty Parlors, General Stores, Bars, Hangouts, and How They Get You through the Day.* New York: Paragon House.

Olshan, Marc A. 1981. "Modernity, the Folk Society, and the Old Order Amish: An Alternative Interpretation." *Rural Sociology,* 46, no. 2(Summer): 297–309.

Ophuls, William. 1977. *Ecology and the Politics of Scarcity.* San Francisco: W. H. Freeman.

Osborne, David. 1990. "Refining State Technology Programs." *Issues in Science and Technology,* 6, no. 4(Summer): 55–61.

Osborne, David, and Ted Gaebler. 1993. *Reinventing Government: How the Entrepreneurial Spirit Is Transforming the Public Sector.* New York: Plume.

Ospina, José. 1987. *Housing Ourselves.* London: Hilary Shipman.

[OTA]. U.S. Congress, Office of Technology Assessment. 1976. *Coastal Effects of Offshore Energy Systems: An Assessment of Oil and Gas Systems, Deepwater Ports, and Nuclear Powerplants off the Coast of New Jersey and Delaware.* Washington, DC: U.S. Government Printing Office.

————. 1982. *Technology and Handicapped People.* OTA-H-179. Washington, DC: U.S. Government Printing Office.

————. 1984. *Computerized Manufacturing Automation: Employment, Education, and the Workplace.* OTA-CIT-235. Washington, DC: U.S. Government Printing Office.

————. 1987. *Science, Technology, and the Constitution—Background Paper.* OTA-BP-CIT-43. Washington, DC: U.S. Government Printing Office.

————. 1988a. *Criminal Justice, New Technologies, and the Constitution.* OTA-CIT-366. Washington, DC: U.S. Government Printing Office.

————. 1988b. *Informing the Nation: Federal Information Dissemination in an*

Electronic Age. OTA-CIT-396. Washington, DC: U.S. Government Printing Office.

———. 1991. *Adolescent Health*. 3 vols. Washington, DC: U.S. Government Printing Office.

———. 1993. *Policy Analysis at OTA: A Staff Assessment*. Washington, DC: OTA, May.

Pacey, Arnold. 1976. *The Maze of Ingenuity: Ideas and Idealism in the Development of Technology*. Cambridge, MA: MIT Press.

———. 1983. *The Culture of Technology*. Cambridge, MA: MIT Press.

Paehlke, Robert. 1990. "Environmental Values and Democracy: The Challenge of the Next Century." In *Environmental Policy in the 1990s: Toward a New Agenda*. Eds. Norman J. Vig and Michael E. Kraft. Washington, DC: Congressional Quarterly Press. Pp. 349–367.

Palca, Joseph. 1989. "AIDS Drug Trials Enter New Age." *Science*, 246(6 Oct.): 19–21.

Park, Peter, Mary Brydon-Miller, Budd Hall, and Ted Jackson, eds. 1993. *Voices of Change: Participatory Research in the United States and Canada*. Westport, CT: Bergin and Garvey.

Patterson, Walter C. 1980. "The Open University Tackles Control of Technology." *Bulletin of the Atomic Scientists*, 36, no. 2(March): 56–57.

Pearlstein, Steven, et al. 1994. "Clinton Puts His Stamp on a 'New Democrat' Budget." *Washington Post National Weekly Edition*, 14–20 Feb., pp. 31–32.

Pelto, Pertti J., and Ludger Müller-Wille. 1987. "Snowmobiles: Technological Revolution in the Arctic." In *Technology and Social Change*. 2nd ed. Eds. H. Russell Bernard and Pertti J. Pelto. Prospect Heights, IL: Waveland Press. Pp. 207–241.

Perrin, Noel. 1980. *Giving up the Gun: Japan's Reversion to the Sword, 1543–1879*. Boulder, CO: Shambhala Press.

Petersen, James C., ed. 1984. *Citizen Participation in Science Policy*. Amherst: University of Massachusetts Press.

Peterson, Jon A. 1982. "Environment and Technology in the Great City Era of American History." *Journal of Urban History*, 8, no. 3(May): 343–354.

Pfaffenberger, Bryan. 1992. "Technological Dramas." *Science, Technology, and Human Values*, 17, no. 3(Summer): 282–312.

Picciotto, Sol. 1988. "The Control of Transnational Capital and the Democratisation of the International State." *Journal of Law and Society*, 15, no. 1(Spring): 58–76.

Piller, Charles. 1991. *The Fail-Safe Society: Community Defiance and the End of American Technological Optimism*. Berkeley and Los Angeles: University of California Press.

Piore, Michael J., and Charles F. Sabel. 1984. *The Second Industrial Divide: Possibilities for Prosperity*. New York: Basic Books.

Polanyi, Michael. 1969. *Knowing and Being: Essays by Michael Polanyi*. Ed. Marjorie Grene. Chicago: University of Chicago Press.

Popper, Karl K. 1966. *The Open Society and Its Enemies*, 2 vols. 5th ed. Princeton, NJ: Princeton University Press.

Porter, Gareth, and Janet Welsh Brown. 1991. *Global Environmental Politics*. Boulder, CO: Westview Press.

Porter, Michael E. 1990. "The Competitive Advantage of Nations." *Harvard Business Review*, 68, no. 2(March–April): 73–93.

Prewitt, Kenneth. 1983. "Scientific Illiteracy and Democratic Theory." *Daedalus*, 112, no. 2(Spring): 49–64.

Primack, Joel, and Frank von Hippel. 1974. *Advice and Dissent: Scientists in the Political Arena*. New York: Basic Books.

Pye, David. 1982. *The Nature and Aesthetics of Design*. New York: Van Nostrand Reinhold.

Rabinow, Paul, and William M. Sullivan, eds. 1979. *Interpretive Social Science: A Reader*. Berkeley and Los Angeles: University of California Press.

Rae, Douglas, Douglas Yates, Jennifer Hochschild, Joseph Morone, and Carol Fessler. 1981. *Equalities*. Cambridge, MA: Harvard University Press.

Ravn, Jørn. n.d. "The Board of Technology and Experience of Technology Assessment." Copenhagen: Danish Board of Technology. MS, 9 pages.

Rawls, John. 1971. *A Theory of Justice*. Cambridge, MA: Harvard University Press.

Rayman, Paula. 1984. "Collective Organization and the National State: The Kibbutz Model." In *Critical Studies in Organization and Bureaucracy*. Eds. Frank Fischer and Carmen Sirianni. Philadelphia: Temple University Press. Pp. 406–420.

Reich, Robert B. 1992. *The Work of Nations*. New York: Vintage Books.

Renner, Michael. 1992. "Creating Sustainable Jobs in Industrial Countries." In *State of the World 1992*. By Lester R. Brown et al. New York: Norton. Pp. 138–154, 233–39.

Rheingold, Howard. 1991. *Virtual Reality*. New York: Simon and Schuster.

Ricoeur, Paul. 1977. "The Model of the Text: Meaningful Action Considered as a Text." In *Understanding and Social Inquiry*. Eds. Fred Dallmayr and Thomas A. McCarthy. Notre Dame, IN: University of Notre Dame Press. Pp. 316–334.

Rifkin, Jeremy. 1983. *Algeny*. New York: Viking Press.

Rip, Arie. 1988. "The Interest of the Netherlands Organisation for Technology Assessment (NOTA) in Studies of Science and Technology." Paper presented at the 4S/EASST Conference, Amsterdam, 16–19 November 1988. Enschede, the Netherlands: University of Twente. MS, 16 pages.

Robertson, John A. 1977. "The Scientist's Right to Research: A Constitutional Analysis." *Southern California Law Review*, 51: 1203–1279.

Rochlin, Gene I. 1986. " 'High-Reliability' Organizations and Technical Change: Some Ethical Problems and Dilemmas." *IEEE Technology and Society Magazine*, 5, no. 3(Sept.): 3–9.

Rogers, Everett M. 1983. *Diffusion of Innovations*. 3rd ed. New York: Free Press.

Rorty, Richard. 1989. *Contingency, Irony, and Solidarity*. Cambridge, UK: Cambridge University Press.

Rose, Mark H. 1988. "Urban Gas and Electric Systems and Social Change, 1900–1940." In *Technology and the Rise of the Networked City in Europe and*

America. Eds. Joel A. Tarr and Gabriel Dupuy. Philadelphia: Temple University Press. Pp. 229–245.

Rose, Mark H., and Joel A. Tarr, eds. 1987. "The City and Technology." *Journal of Urban History*, 14, no. 1(Nov.): 3–139.

Rose-Ackerman, Susan. 1992. *Rethinking the Progressive Agenda: The Reform of the American Regulatory State.* New York: Free Press.

Rosen, Christine Meisner. 1986. "Infrastructural Improvement in Nineteenth-Century Cities: A Conceptual Framework and Cases." *Journal of Urban History*, 12, no. 3(May): 211–256.

Rosenberg, Nathan. 1979. "Technology, Economy, and Values." In *The History and Philosophy of Technology.* Eds. George Bugliarello and Dean B. Doner. Chicago: University of Chicago Press. Pp. 81–111.

———. 1994. *Exploring the Black Box: Technology, Economics, and History.* Cambridge, UK: Cambridge University Press.

Rosenthal, Debra. 1990. *At the Heart of the Bomb: The Dangerous Allure of Weapons Work.* Reading, MA: Addison-Wesley.

Rosner, Menachem. 1983. *Democracy, Equality, and Change: The Kibbutz and Social Theory.* Darby, PA: Norwood Editions.

Rothschild, Joyce, and J. Allen Whitt. 1986. *The Cooperative Workplace: Potentials and Dilemmas of Organizational Democracy and Participation.* Cambridge, UK: Cambridge University Press.

Rousseau, Jean-Jacques. 1968. *The Social Contract.* Trans. Maurice Cranston. New York: Penguin Books.

Rudofsky, Bernard. 1964. *Architecture without Architects: A Short Introduction to Non-Pedigreed Architecture.* Garden City, NY: Doubleday.

Sabbah, Françoise. 1985. "The New Media." In *High Technology, Space, and Society.* Ed. Manuel Castells. Beverly Hills, CA: Sage. Pp. 210–224.

Sabel, Charles F. 1982. *Work and Politics: The Division of Labor in Industry.* Cambridge, UK: Cambridge University Press.

Sabel, Charles F., and Jonathan Zeitlin. 1985. "Historical Alternatives to Mass Production: Politics, Markets and Technology in Nineteenth-Century Industrialization." *Past and Present*, no. 108(Aug.), pp. 133–176.

Sachs, Wolfgang. 1991. "Environment and Development: The Story of a Dangerous Liaison." *Ecologist*, 21, no. 6(Nov.–Dec.): 252–257.

Sagoff, Mark. 1988. *The Economy of the Earth: Philosophy, Law, and the Environment.* Cambridge, UK: Cambridge University Press.

Samuelson, Paul A. 1966. *The Collected Scientific Papers of Paul A. Samuelson.* Vol. II. Ed. Joseph E. Stiglitz. Cambridge, MA: MIT Press.

Sandberg, Åke. 1993. "Volvo Human-Centered Work Organization—The End of the Road?" *New Technology, Work, and Employment*, 8, no. 2(Sept.): 83–87.

Sandberg, Åke, Gunnar Broms, Arne Grip, Lars Sundström, Jesper Stren, and Peter Ullmark. 1992. *Technological Change and Co-Determination in Sweden.* Philadelphia: Temple University Press.

Sandel, Michael J. 1982. *Liberalism and the Limits of Justice.* Cambridge, UK: Cambridge University Press.

Schäfer, Wolf. 1982. "Collective Thinking from Below: Early Working-Class Thought Reconsidered." *Dialectical Anthropology*, 6, no. 3(March): 193–214.

Scheuer, Jeffrey, and Richard Sclove. 1992. "Ways to Reach Voters Changing." *Houston Post*, 23 Aug., pp. C1, C6.

Schiller, Herbert L. 1993. "Transnational Media: Creating Consumers Worldwide." *Journal of International Affairs*, 47, no. 1(Summer): 47–58.

Schmitt, Eric. 1993. "Arms for Job and Country." *New York Times*, 2 Sept., pp. A1, A18.

Schneider, Keith. 1990. "Wisconsin Temporarily Banning Gene-Engineered Drug for Cows." *New York Times*, 28 April, pp. 1, 11.

Schor, Juliet B. 1992. *The Overworked American: The Unexpected Decline of Leisure*. New York: Basic Books.

Schot, Johan W. 1992. "Constructive Technology Assessment and Technology Dynamics: The Case of Clean Technologies." *Science, Technology, and Human Values*, 17, no. 1(Winter): 36–56.

Schot, Johan W., Remco Hoogma, and Boelie Elzen. 1994. "Strategies for Shifting Technological Trajectories." *Futures*, 26, no. 10(Dec.): 1060–1076.

Schrank, Robert. 1978. *Ten Thousand Working Days*. Cambridge, MA: MIT Press.

_____, ed. 1979. *American Workers Abroad: A Report to the Ford Foundation*. Cambridge, MA: MIT Press.

Schudson, Michael. 1986. *Advertising, the Uneasy Persuasion: Its Dubious Impact on American Society*. New York: Basic Books.

Schuler, Douglas, and Aki Namioka, eds. 1993. *Participatory Design: Principles and Practices*. Hillsdale, NJ: Lawrence Erlbaum Associates.

Schumacher, E. F. 1973. *Small Is Beautiful: Economics as if People Mattered*. New York: Perennial Library.

Schutz, Alfred. 1946. "The Well-Informed Citizen: An Essay on the Social Distribution of Knowledge." *Social Research*, 13, no. 4(Dec.): 463–478.

Schwartz, Adina. 1982. "Meaningful Work." *Ethics*, 92, no. 4(July): 634–646.

Schwartz, Joel D. 1984. "Participation and Multisubjective Understanding: An Interpretivist Approach to the Study of Political Participation." *Journal of Politics*, 46, no. 4(Nov.): 1117–1141.

Schwarz, Michiel, and Michael Thompson. 1990. *Divided We Stand: Redefining Politics, Technology and Social Choice*. Philadelphia: University of Pennsylvania Press.

Sclove, Richard E. 1982. "Decision-Making in a Democracy." *Bulletin of the Atomic Scientists*, 38, no. 5(May): 44–49.

_____. 1983. "Energy Policy and Democratic Theory." In *Uncertain Power: The Struggle for a National Energy Policy*. Ed. Dorothy S. Zinberg. New York: Pergamon Press. Pp. 37–65.

_____. 1989. "From Alchemy to Atomic War: Frederick Soddy's 'Technology Assessment' of Atomic Energy, 1900–1915." *Science, Technology, and Human Values*, 14, no. 2(Spring): 163–194.

_____. 1992. "The Nuts and Bolts of Democracy: Democratic Theory and Technological Design." In *Democracy in a Technological Society.* Philosophy and Technology, vol. 9. Ed. Langdon Winner. Dordrecht: Kluwer Academic Publishers. Pp. 139–157.

_____. 1994a. "Democratizing Technology." *Chronicle of Higher Education,* 40, no. 19 (12 Jan): pp. B1–B2.

_____. 1994b. *Technology, Society, and Democracy: New Problems and Opportunities.* A Report to the General Program of the MacArthur Foundation. Chicago: John D. and Catherine T. MacArthur Foundation, May.

_____. 1995. "Putting Science to Work in Communities." *Chronicle of Higher Education,* 41, no. 29 (31 March): B1–B3.

Sclove, Richard, and Jeffrey Scheuer. 1994. "The Ghost in the Modem: For Architects of the Info-Highway, Some Lessons from the Concrete Interstate." *Washington Post,* Outlook Section, 29 May, p. C3.

Sen, Amartya. 1985. "The Moral Standing of the Market." In *Ethics and Economics.* Eds. Ellen Frankel Paul, Jeffrey Paul, and Fred D. Miller, Jr. Oxford: Basil Blackwell. Pp. 1–19.

Sesser, Stan. 1992. "A Reporter at Large: A Nation of Contradictions." *New Yorker,* 13 Jan., pp. 37–68.

Sewell, John W. 1990. "Toward North-South Partnership." *Issues in Science and Technology,* 6, no. 3(Spring): 47–52.

Shaiken, Harley. 1985. "The Automated Factory: The View from the Shop Floor." *Technology Review,* 88, no. 1(Jan.): 17–24.

Shavelson, Jeff. 1990. *A Third Way, a Sourcebook: Innovations in Community-Owned Enterprise.* Washington, DC: National Center for Economic Alternatives.

Shaw, Bradley T. 1987. "Technology and Military Criteria: Broadening the Theory of Innovation." *Technological Forecasting and Social Change,* 31: 239–256.

Shiva, Vandana. 1991. *The Violence of the Green Revolution: Third World Agriculture, Ecology and Politics.* London: Zed Books.

Shrader-Frechette, Kristin. 1983. "Technology Assessment and the Problem of Quantification." In *Philosophy and Technology.* Eds. Paul T. Durbin and Friedrich Rapp. Dordrecht: D. Reidel. Pp. 151–164.

_____. 1985. *Science Policy, Ethics, and Economic Methodology: Some Problems of Technology Assessment and Environmental-Impact Analysis.* Dordrecht: D. Reidel.

Shuman, Michael. 1994. "GATTzilla v. Communities." *Cornell International Law Journal,* 27(Sept.): 101–124.

Shuman, Michael, and Hal Harvey. 1993. *Security without War: A Post–Cold War Foreign Policy.* Boulder, CO: Westview Press.

Shuman, Michael, and Julia Sweig, eds. 1993. *Technology for the Common Good.* Washington, DC: Institute for Policy Studies.

Shweder, Richard A., and Robert A. LeVine, eds. 1984. *Culture Theory: Essays on Mind, Self, and Emotion.* Cambridge, UK: Cambridge University Press.

Siefert, Marsha, George Gerbner, and Janice Fisher, eds. 1989. *The Information Gap: How Computers and Other New Communication Technologies Affect the Social Distribution of Power.* Oxford: Oxford University Press.

Sirianni, Carmen J. 1981. "Production and Power in a Classless Society: A Critical Analysis of the Utopian Dimensions of Marxist Theory." *Socialist Review*, 11, no. 5(Sept.-Oct.): 33–82.

_____. 1983. "The Council Model of Decentralized Planning: A Critical Analysis." In *Socialist Visions*. Ed. Steve Rosskam Shalom. Boston: South End Press. Pp. 279–286.

_____. 1984. "Participation, Opportunity, and Equality: Toward a Pluralist Organizational Model." In *Critical Studies in Organization and Bureaucracy.* Eds. Frank Fischer and Carmen Sirianni. Philadelphia: Temple University Press. Pp. 482–503.

_____. 1988. "Self-Management of Time: A Democratic Alternative." *Socialist Review*, 18, no. 4(Oct.-Dec.): 5–56.

Smart, J. J. C., and Bernard Williams. 1973. *Utilitarianism: For and Against.* Cambridge, UK: Cambridge University Press.

Smit, Wim A., John Grin, and Lev Voronkov, eds. 1992. *Military Technological Innovation and Stability in a Changing World: Politically Assessing and Influencing Weapon Innovation and Military Research and Development.* Amsterdam: VU University Press.

Smith, Adam. 1937. *An Inquiry into the Nature and Causes of the Wealth of Nations.* New York: Modern Library.

Smith, Cyril Stanley. 1970. "Art, Technology, and Science: Notes on Their Historical Interaction." *Technology and Culture*, 11, no. 4(Oct.): 493–549.

Smith, Merritt Roe, ed. 1985. *Military Enterprise and Technological Change.* Cambridge, MA: MIT Press.

Sommer, Robert. 1969. *Personal Space: The Behavioral Basis of Design.* Englewood Cliffs, NJ: Prentice-Hall.

_____. 1983. *Social Design: Creating Buildings with People in Mind.* Englewood Cliffs, NJ: Prentice-Hall.

Sorkin, Michael, ed. 1992. *Variations on a Theme Park: The New American City and the End of Public Space.* New York: Noonday Press.

Spivack, Mayer. 1969. "The Political Collapse of a Playground." *Landscape Architecture*, 59, no. 4(July): 288–292.

Stefanik, Nancy. 1993. "Sustainable Dialogue/Sustainable Development: Developing Planetary Consciousness via Electronic Democracy." In *Global Visions: Beyond the New World Order.* Eds. Jeremy Brecher et al. Boston: South End Press. Pp. 263–272.

Stephens, Carlene. 1989. " 'The Most Reliable Time': William Bond, the New England Railroads, and Time-Awareness in 19th-Century America." *Technology and Culture*, 30, no. 1(Jan.): 1–24.

Stigler, George J., and Gary S. Becker. 1977. "De Gustibus non Est Disputandum." *American Economic Review*, 67, no. 2: 76–90.

Stirling, Andrew. 1993. "Environmental Valuation: How Much Is the Emperor Wearing?" *Ecologist*, 23, no. 3(May–June): 97–103.

Stokey, Edith, and Richard Zeckhauser. 1978. *A Primer for Policy Analysis*. New York: W. W. Norton.

Stoltzfus, Victor. 1973. "Amish Agriculture: Adaptive Strategies for Economic Survival of Community Life." *Rural Sociology*, 38, no. 2(Summer): 196–206.

Strasser, Susan. 1989. *Satisfaction Guaranteed: The Making of the American Mass Market*. New York: Pantheon Books.

Summerton, Jane, ed. 1994. *Changing Large Technical Systems*. Summertown, Oxford, UK: Westview.

Sun, Marjorie. 1984. "Weighing the Social Costs of Innovation." *Science*, 223(30 March): 1368–1369.

Sutherland, George H. 1993. "Statement of George H. Sutherland before the Committee on Labor and Human Resources, U.S. Senate." Cleveland: Great Lakes Manufacturing Technology Center. MS, 5 pages.

Suzuki, Shunryu. 1970. *Zen Mind, Beginner's Mind*. New York: Weatherhill.

Tarr, Joel A. 1988. "Sewerage and the Development of the Networked City in the United States, 1850–1930." In *Technology and the Rise of the Networked City in Europe and America*. Eds. Joel A. Tarr and Gabriel Dupuy. Philadelphia: Temple University Press. Pp. 159–185.

Taylor, Michael. 1982. *Community, Anarchy and Liberty*. Cambridge, UK: Cambridge University Press.

Taylor, S. Martin. 1989. "Community Exclusion of the Mentally Ill." In *The Power of Geography: How Territory Shapes Social Life*. Eds. Jennifer Wolch and Michael Dear. Boston: Unwin Hyman. Pp. 316–330.

Technology and Democracy: The Use and Impact of Technology Assessment in Europe. n.d. Proceedings of the 3rd European Congress on Technology Assessment, Copenhagen, 4–7 November 1992. 2 vols. Copenhagen: TeknologiNaevnet.

Technology for Economic Growth: President's Progress Report. 1993. Washington, DC: The White House, Nov.

Texas Department of Commerce. 1992. *Comparative Evaluation of Marketplace Systems*. San Antonio: Texas Dept. of Commerce, Nov.

Thomas, Robert J. 1994. *What Machines Can't Do: Politics and Technology in the Industrial Enterprise*. Berkeley and Los Angeles: University of California Press.

Thompson, Dennis F. 1983. "Bureaucracy and Democracy." In *Democratic Theory and Practice*. Ed. Graeme Duncan. Cambridge, UK: Cambridge University Press. Pp. 235–250.

Thompson, William Irwin. 1971. *At the Edge of History: Speculations on the Transformation of Culture*. New York: Harper/Colophon.

Thrupp, Lori Ann. 1989. "Legitimizing Local Knowledge: From Displacement to Empowerment for Third World People." *Agriculture and Human Values*, 6, no. 3(Summer): 13–24.

Tickner, J. Ann. 1987. *Self-Reliance versus Power Politics: The American and Indian Experiences in Building Nation States*. New York: Columbia University Press.

Todd, Edmund N. 1987. "A Tale of Three Cities: Electrification and the Structure of Choice in the Ruhr, 1886–1900." *Social Studies of Science*, 17, no. 3(Aug.): 387–412.

Todd, John, and Nancy Jack Todd. 1984. *Bioshelters, Ocean Arks, City Farming: Ecology as the Basis of Design*. San Francisco: Sierra Club Books.

Tribe, Laurence H. 1973. "Technology Assessment and the Fourth Discontinuity: The Limits of Instrumental Rationality." *Southern California Law Review*, 46, no. 3(June): 617–660.

Tripp, Alice. 1980. "Powerline Assaults the Prairie." *Science for the People*, 12, no. 5(Sept.–Oct.): 19–21, 33.

Turkle, Sherry. 1984. *The Second Self: Computers and the Human Spirit*. New York: Simon and Schuster.

Turnbull, Colin M. 1972. *The Mountain People*. New York: Simon and Schuster.

Uchitelle, Louis. 1994. "Job Losses Don't Let Up Even as Hard Times Ease." *New York Times*, 22 March, pp. A1, D5.

Ullmann, John E. 1990. "Shaking Off Cold War Ideology: Intellectual and Political Change in a Demilitarized Society." Briefing Paper no. 8. Washington, DC: National Commission for Economic Conversion and Disarmament.

Unger, Roberto Mangabeira. 1975. *Knowledge and Politics*. New York: Free Press.

———. 1987. *Social Theory: Its Situation and Its Task*. Cambridge, UK: Cambridge University Press.

van Creveld, Martin. 1989. *Technology and War: From 2000 B.C. to the Present*. New York: Free Press.

van Dam, Laura, and Robert Howard. 1988. "Life at the OTA." *Technology Review*, 91, no. 7(Oct.): 46–51.

van den Broecke, Marcel P.R. 1993. "Science Advice Centres in the Netherlands: A Bird's Eye View." Paper prepared for Public Information on Science, a British Association–Rutherford Trust Conference, 8–10 April, 1993, Edinburgh, United Kingdom. MS, 6 pages.

Vig, Norman J. 1992. "Parliamentary Technology Assessment in Europe: Comparative Evolution." *Impact Assessment Bulletin*, 10, no. 4: 3–24.

Vogel, Carl. 1993. "Money Makers: Turning Community Talent into Local Currency." *Neighborhood Works*, 16, no. 4(Aug.–Sept.): 14–15.

Wad, Atul, ed. 1988. *Science, Technology, and Development*. Boulder, CO: Westview Press.

Wagner, Roy. 1981. *The Invention of Culture*. Rev. ed. Chicago: University of Chicago Press.

Wainwright, Hilary, and Dave Elliott. 1982. *The Lucas Plan: A New Trade Unionism in the Making?*. London: Allison and Busby.

Wajcman, Judy. 1991. *Feminism Confronts Technology*. University Park: Pennsylvania State University Press.

Walton, Richard E., and Michael E. Gafney. 1991. "Research, Action, and Participation: The Merchant Shipping Case." In *Participatory Action Research*. Ed. William Foote Whyte. Newbury Park, CA: Sage. Pp. 99–126.

Ward, Colin. 1978. *The Child in the City*. New York: Pantheon Books.

_____. 1983. *Housing: An Anarchist Approach*. 2nd ed. London: Freedom Press.

Weart, Spencer R. 1988. *Nuclear Fear: A History of Images*. Cambridge, MA: Harvard University Press.

Webster, Frank, and Kevin Robins. 1989. "Plan and Control: Towards a Cultural History of the Information Society." *Theory and Society*, 18, no. 3(May): 323–351.

Weizenbaum, Joseph. 1976. *Computer Power and Human Reason: From Judgment to Calculation*. San Francisco: W. H. Freeman.

Wharton, Donald. 1980. "Designing with Users: Developing the Lorena Stove, Guatemala." In *Experiences in Appropriate Technology*. Ed. Robert J. Mitchell. Ottawa: Canadian Hunger Foundation. Pp. 21–26.

White, Lynn, Jr. 1962. *Medieval Technology and Social Change*. Oxford: Oxford University Press.

_____. 1974. "Technology Assessment from the Stance of a Medieval Historian." *Technological Forecasting and Social Change*, 6: 359–369.

Whyte, William Foote. 1990. "New Ways of Organizing Industrial Work." In *To Govern a Changing Society: Constitutionalism and the Challenge of New Technology*. Ed. Robert S. Peck. Washington, DC: Smithsonian Institution Press. Pp. 169–183.

_____, ed. 1991. *Participatory Action Research*. Newbury Park, CA: Sage.

Whyte, William Foote, and Joseph R. Blasi. 1984. "Worker Ownership, Participation, and Control: Toward a Theoretical Model." In *Critical Studies in Organization and Bureaucracy*. Eds. Frank Fischer and Carmen Sirianni. Philadelphia: Temple University Press. Pp. 377–405.

Whyte, William Foote, and Kathleen King Whyte. 1988. *Making Mondragon: The Growth and Dynamics of the Worker Cooperative Complex*. Ithaca, NY: ILR Press, Cornell University.

Whyte, William H. 1988. *City: Rediscovering the Center*. New York: Anchor Books.

Williams, Karel, Colin Haslam, John Williams, and Tony Cutler with Andy Adcroft and Sukhdev Johal. 1992. "Against Lean Production." *Economy and Society*, 21, no. 3(Aug.): 321–354.

Williams, Robert H., Eric D. Larson, and Marc H. Ross. 1987. "Materials, Affluence, and Industrial Energy Use." *Annual Review of Energy*, 12: 99–144.

Williams, Robert H., and Eric D. Larson. 1989. "Expanding Roles for Gas Turbines in Power Generation." In *Electricity: Efficient End-Use and New*

Generation Technologies, and Their Planning Implications. Eds. Thomas B. Johansson et al. Lund, Sweden: Lund University Press. Pp. 503–553.

Wills, Garry. 1978. *Inventing America: Jefferson's Declaration of Independence*. Garden City, NY: Doubleday.

Wilson, William Julius. 1987. *The Truly Disadvantaged: The Inner City, the Underclass, and Public Policy*. Chicago: University of Chicago Press.

Winn, Marie. 1985. *The Plug-in Drug: Television, Children, and the Family*. Rev. ed. New York: Penguin Books.

Winner, Langdon. 1977. *Autonomous Technology: Technics-out-of-Control as a Theme in Political Thought*. Cambridge, MA: MIT Press.

———. 1986. *The Whale and the Reactor: A Search for Limits in an Age of High Technology*. Chicago: University of Chicago Press.

Wisbrock, Rollie. 1990. "Oregon Marketplace Revisited." *Economic Development Commentary*, 14, no. 3(Fall): 8–12.

Wolf, Eric R. 1982. *Europe and the People without History*. Berkeley and Los Angeles: University of California Press.

Wolf, Naomi. 1991. *The Beauty Myth: How Images of Beauty Are Used against Women*. New York: William Morrow.

Wolfe, Alan. 1989. *Whose Keeper?: Social Science and Moral Obligation*. Berkeley and Los Angeles: University of California Press.

Wolff, Robert Paul, ed. 1983. "Philosophy and Economics." *Philosophical Forum*, 14, nos. 3–4(Spring–Summer): i–vi, 211–403.

Wolkomir, Richard. 1985. "A Playful Designer Who Believes That Kids Know Best." *Smithsonian*, Aug., pp. 106–115.

Wright, Barbara Drygulski, ed. 1987. *Women, Work, and Technology: Transformations*. Ann Arbor: University of Michigan Press.

Wynne, Brian. 1988. "Unruly Technology: Practical Rules, Impractical Discourses, and Public Understanding." *Social Studies of Science*, 18: 147–167.

Yanagi, Soetsu. 1978. *The Unknown Craftsman: A Japanese Insight into Beauty*. Adapted by Bernard Leach. Tokyo: Kodansha International.

Yates, JoAnne. 1989. *Control through Communication: The Rise of System in American Management*. Baltimore: Johns Hopkins University Press.

Young, Iris Marion. 1989. "Polity and Group Difference: A Critique of the Ideal of Universal Citizenship." *Ethics*, 99, no. 2(Jan.): 250–274.

———. 1990. *Justice and the Politics of Difference*. Princeton, NJ: Princeton University Press.

Zaal, Rolf, and Loet Leydesdorff. 1987. "Amsterdam Science Shop and Its Influence on University Research: The Effects of Ten Years of Dealing with Non-Academic Questions." *Science and Public Policy*, 14, no. 6(Dec.): 310–316.

Zimbalist, Andrew. 1975. "The Limits of Work Humanization." *Review of Radical Political Economy*, 7, no. 2: 50–59.

———, ed. 1979. *Case Studies on the Labor Process*. New York: Monthly Review Press.

Zimmerman, Jan. 1986. *Once upon the Future: A Woman's Guide to Tomorrow's Technology*. New York: Pandora.

Zimmerman, Michael. 1979. "Heidegger and Marcuse: Technology as Ideology." *Research in Philosophy and Technology*, 2: 245–261.

Zoglin, Richard. 1989. "Subversion by Cassette." *Time*, 11 Sept., p. 80.

Zuboff, Shoshana. 1988. *In the Age of the Smart Machine: The Future of Work and Power*. New York: Basic Books.

INDEX

Note: Page numbers in *italic* refer to figures. Page numbers with *n* refer to notes.